T0178082

# A random walk in science

## The Compiler

Robert L Weber (deceased) drew on long years of experience as an educator, author and editor to illustrate the humour and humanism in science to prove that the subject can be entertaining as well as enlightening. He was Associate Professor of physics at The Pennsylvania State University, and the author of more than a dozen books. He served on the boards of scientific and scholarly publications and regularly reviewed books for a wide range of scientific publications. Dr Weber received his BA at Yale University and his PhD at The Pennsylvania State University. Sadly, he died in 1997.

## The Editor

Professor Eric Mendoza became interested in education while at the University of Manchester. He was mainly responsible for reforming the physics syllabus at Manchester and later at the University College of North Wales, and he is now furthering his interest in education at the Israel Science Teaching Centre at the Hebrew University, Jerusalem.

Robert Weber continued to gather humorous stories, anecdotes, verse and cartoons, producing two further anthologies *More Random Walks in Science* (Institute of Physics Publishing, 1982) and *Science with a Smile* (Institute of Physics Publishing, 1992).

*'I think I can guarantee that virtually every reader will find something to tickle his or her funny bone within these volumes'*
Physics Today

# A random walk in science

An anthology compiled
by the late R L Weber (1913–1997)

Edited by E Mendoza

With a Foreword by William Cooper

CRC Press
Taylor & Francis Group
Boca Raton London New York

CRC Press is an imprint of the
Taylor & Francis Group, an **informa** business

Originally published by The Institute of Physics

First published 1973 (hbk)

Published as a paperback 1999

Published 2023 by CRC Press
Taylor & Francis Group
6000 Broken Sound Parkway NW, Suite 300
Boca Raton, FL 33487-2742

ISBN 13: 978-0-7503-0649-2 (pbk)

Visit the Taylor & Francis Web site at
http://www.taylorandfrancis.com

and the CRC Press Web site at
http://www.crcpress.com

*Library of Congress Cataloging-in-Publication Data are available*

*British Library Cataloguing-in-Publication Data*

A catalogue record for this book is available from the British Library.

Designed by Bernard Crossland
Text set in 12 pt Barbou and 11 pt Times New Roman
Cover Design: Jeremy Stephens

# Foreword

WILLIAM COOPER

I must say, clever men are fun. It struck me afresh, just reading a sample The Institute of Physics sent me in advance of contributions to *A random walk in science*. (Naturally the sample was representative.) Fun—that's not, necessarily, to say funny; though some of the contributions are very funny. Fun, as I'm defining it for the moment in my own lexicon, arises from a play of intellectual high spirits, or high intellectual spirits. (I'm not fussy about which order the words come in, being neither Wittgensteinian about what can and can't be said, nor French about linguistic precision—lots of things worth saying can only be said loosely.)

In fact spiritedly high intellect also goes for what I'm trying to get at. With high intelligence there's nearly always an overflow of intellectual energy, free energy available for vitalizing any old topic that comes up, or, better still, for incarnating new ones out of the empyrean. It's the play of this free intellectual energy that makes the person who generates it fun to read, fun to be with. Perhaps I ought to confess, now, that my private subtitle for this volume is 'Physicists At Play'.

So while readers of *A random walk in science* are being promised fun, the contributors find themselves being called clever. Well, there's something in that. It has always seemed clear to me that level of intelligence is much more decisive in the sorting-out of scientists than it is in the sorting-out of, say, writers. (I've chosen writers for comparison with scientists so as to keep sight of the 'creative' element in what they both do.) My general impression, for instance in moving between a group of scientists and a comparable group of writers, comparable in distinction of talent and reputation, is of a drop in the average IQ. To take a specific case: I should have thought you simply couldn't be a first-rate physicist without a first-rate intellectual equipment; whereas you *can* be a first-rate novelist—quite a few have been.

Such as who? you ask. Trying to avoid the most obvious dangers in the present circumstances, by going to the top flight in distinction and choosing a scientist who's not a physicist and a novelist who's not alive, I suggest juxtaposing Jacques Monod and D H Lawrence. (I *know* that the possession of highest intellect is not what we primarily require of a novelist; that's *not* what this argument's about.) I feel that by any of the criteria we normally accept for judging intellectual power and range, Lawrence, though he's pretty well bound to be placed in the top flight of novelists, simply has to come in a flight below Monod as a mind. (It's particularly amusing to imagine the rage of Lawrencians at the demotion of their prophet as a mind—when the message which they receive from him with such reverence and passion is patently *anti*-mind!) And if one comes down the flights from the top, I think a similar juxtaposing on almost any of them would most frequently give the edge to the scientist, certainly to the practitioner of the 'exact' sciences.

Having then fulfilled the two prime requirements for a Foreword-writer—(i) to promise the readers and (ii) to flatter the authors—I can get on with saying something more about the contributions. For instance, what sort of fun is it that characterizes physicists at play? It's the fun of playing tricks with conceptual thought—misapplying concepts, parodying them, standing them on their heads. I have a special weakness, myself, for tricks being played with the concepts of mathematics and symbolic logic—'A Contribution to the Mathematical Theory of Big Game Hunting', which shows how to trap a lion in the Sahara sheerly by manipulating ideas, suits me excellently.

But the whole book is far from being confined to playing with mathematics and symbolic logic. There's a selection of in-jokes by physicists at their most worldly—in-jokes that can readily be understood by non-physicists, since a lot of them are making sarcastic fun of how the world works, on which physicists cast a very beady eye as a result of having to cope with it—where 'cope' usually means 'crash through it in order to get some physics done.' O & M wreaking their uncomprehending will at the Festival Hall; 'Why we should go to the Moon' (because 'the world is running dangerously short of unprocessed data'); a 'Proposal for a Coal Reactor'. And jokes at their own expense—the gamesmanship of physicists; cynical glossaries of the professional terms they use, and so on. Very funny and, indeed, very worldly.

Yet this fun is still essentially located more in the realm of the conceptual than of the human. (If you asked me now to explain in one sentence what I mean by the 'human', I should say it had something to do with seeing the fun—and the pathos, as well—in a single fellow mortal's being wholly and sheerly *himself*.) And worldliness, when you come to think about it, incorporates a high degree of conceptualizing, of abstracting from general human behaviour within narrow, if amusing, terms of reference. So the expression of physicists at play hangs together quite remarkably. *A random walk in science* keeps one startlingly within a perimeter, a perimeter within which a set of clever men are having a high old time with rational concepts. Their high spirits and confidence are particularly startling to anyone who spends much time outside the perimeter, especially in the part of the culture which is occupied with the arts.

Why is it startling? What is it that enables a set of clever men to live way out there, having a high old intellectual time, on their own? I can only put forward a personal interpretation—at the risk of provoking rage on another front. Let me put it this way: it's easy to say what's inside the perimeter and it's pretty stunning, at that. What is *not* inside it, as it strikes me, is what I should call a deep sense of the darker side of existence, of the tragic nature of the single human being's fate—and, in this context, all that hinges on a sense of how slight, how desperately slight is the hold of rationality on the way we behave.

There are two things I don't mean by that. The first is that physicists don't have a sense of cosmic danger: they do. Once upon a time, The Bomb: now, Ecological Disaster. But a sense of cosmic danger is a totally different thing from a tragic sense of life. The second is that physicists are unaware of irrationality in the individual behaviour of other men and even, at a pinch, of themselves: they are—but in the impatient, exasperated manner of men who have not comprehended that irrationality is our basic natural state.

They recognize that the crucial step on the way to scientific discovery is not rational, but intuitive. Of course. But the scientific discipline teaches one how to evaluate one's intuitions. 'The student of physics has his intuition violated so repeatedly,' writes one of the contributors, with a sort of careless starkness, 'that he comes to accept it as a routine experience.' I take it that all physicists would more or less agree with him. I wonder if they have any intimations of the growing proportion of people in the world now, certainly in the culture we ourselves are living in, who would regard that statement as arising from a view of life which to them is anathema? The devaluation of intuition by mind—evil.

*A random walk in science* begins with a challenge, at once playful in expression and sound at heart, about the Two Cultures. It recognizes the polarization that has taken place, and suggests that it would have been less likely to have taken place round scientific and non-scientific elements in the culture—or having done so, it would be more likely to disappear—if we English had used the word (and the idea of) 'science' broadly to include *all* scholarship, as the Dutch use the word '*wetenschappen*'. It's an amusing idea. But if we used 'science' as he suggests, we should dilute the meaning of the word and have to find a new one to signify what we currently call science. What's more, the Two Cultures polarization happens unfortunately to be just as serious in Holland, anyway.

On the other hand, the idea jives unexpectedly with the argument I'm leading up to. The polarization into the Two Cultures exists; but in my view the form in which it is now manifesting itself is deeper and more alarming than appeared when the poles were seen to be science and non-science. They are now manifesting themselves in a form that shows our situation to be more grave than it would be if the poles were even *wetenschappen* and non-*wetenschappen*. They are mind and anti-mind.

The situation is not Alexandrian, because history doesn't happen twice in the same form; but to think about ancient Alexandria and now is deeply disturbing. In the earlier culture they had marvellous science going on, within its perimeter scientists in high spirits and high confidence; and outside . . . a lapse into complex and arcane fatuity. What do we have now? Excellent science and technology, its practitioners within its perimeter sparkling with high spirits and confidence, living by *mind*; and elsewhere . . . lapse into the fatuity of headless exaltation of the instinctual life, the irrational life—or, to use the current terminology, the 'authentic' life—*anti-mind*.

Lawrence was devoting his art to it fifty years ago. Things have moved on since then. In the present we have, for example, the turning away from learning history, because knowing what happened in the past inhibits one from acting according to instinct now; the regarding of a schizophrenic's madness as his sanity—to live with him *we* must enter *it*; the idiot reverence for drug-experiences, or any other experiences, that 'blow' the mind. And so on, and on.

Thus I summarize my argument. Only men who have a sense of the darker side of human existence, who know in their bones how slight is the grasp of rationality on the instinctive forces that drive us and have intimations of the sterile fatuity that would ensue from being overwhelmed by them—only such men can truly cope with the danger that faces the intellectual world. Reading *A random walk in science* I was entertained, pleased, stimulated, roused to admiration—and troubled. Physicists at play. Are they unconscious of their fate?

# Contents

Contents

xii

# Introduction

It is sad that it should seem necessary today to rescue scientists from the unattractive stereotypes and caricatures with which they are encumbered. Physics, the basic science, seems most in need of humanizing. Older philosophies of science pretended that physics proceeds from certainty to certainty through the performance of critical experiments unambiguously interpreted. This created the impression that physicists themselves have no room for doubt, that they have no emotions and no time for laughter—in short, that they are inhuman.

Much of the misunderstanding of scientists and how they work is due to the standard format of articles in scientific journals. With their terse accounts of successful experiments and well-supported conclusions they show little of the untidy nature of research at the frontiers of knowledge. In self defence, there has grown up a derisive, sometimes cynical attitude of self criticism by scientists, a subculture which transcends geographical and political barriers. Experimenters' gibes at the uselessness of theoreticians, glossaries of the real meanings behind well-worn phrases, disillusion at the corrupting effect of the vast sums of money lavished on government research laboratories, can be found in articles from Russia or America, Britain or continental Europe. On the other hand Rutherford's sensitivity to Nature's whispers, Boltzmann's sense of the sublime in Maxwell's work, or poor William Crabtree's emotion on seeing the transit of Venus, these are attitudes and feelings which every scientist knows are at the centre of scientific research. They rarely show through the language of our reports.

A flourishing underground press has grown up in science. A typical journal is the *Worm Runner's Digest*. 'It started,' says Dr J V McConnell, as 'my own personal joke on the Scientific Establishment although it has turned out to be more of a joke on me. I've lost grants because of the *Digest* . . .'. After twelve years of uninhibited life, the *Digest* is published in two parts. The front half records bona fide research under an acceptable title, *The Journal of Biological Research*; it is noticed in *Psychological Abstracts*, *Biological Abstracts*, and *Chemical Abstracts*. But the second half of the *Digest* remains 'the *Playboy* of the scientific world,' its pages printed upside down to help distinguish fact from fantasy. It is the house organ of an anti-Scientific movement. McConnell's conviction is that 'most of what is wrong with science these days can be traced to the fact that scientists are willing to make objective and dispassionate studies of any natural phenomen at all—except their own scientific behaviour. We know considerably more about flatworms than we do about people who study flatworms. The Establishment never questions its own motives; the true humorist always does.'

In this book I have drawn heavily on such journals and on other informal writings by scientists. It is a collection of comments, both lighthearted and serious, by scientists. They reveal their intensely human ambitions, frustrations and elation; they record some changing attitudes within science and mirror the interactions of science with society.

I hope you find as much pleasure in reading these pages as I did in assembling them.

Professor Eric Mendoza, who kindly consented to serve as The Institute of Physics' Honorary Editor for this book, has been an enthusiastic and careful editor and has brought additional items to the collection. It has been a pleasure to work with him, though at a distance; I express my gratitude for his substantial help.

ROBERT L WEBER

This anthology started life as a collection of jokes about physics. Physicists, thought Professor Weber, took themselves too seriously and would benefit from the opportunity to laugh at themselves. But it was not long before he added another more serious ingredient and broadened the scope to include other subjects close to physics. The manuscript came to be entitled 'Humour and Humanism in Science' and it was in this form that it was submitted to The Institute of Physics. It seemed to me, however, that a collection overwhelmingly drawn from the twentieth century lacked those deeper notes—the graver modes, Rayleigh would have called them—with which physics, with its long and turbulent history, so resonates. The character of the book gradually changed as many cynical wisecracks from today's whizz kids gave place to more measured pronouncements from the giants of our history, and the more obscure in-jokes were discarded in favour of dramas and tragedies from the past.

This is not a scholarly book; it has been arranged for dipping into, for casual reading, and many of the articles have been condensed. To that end, it has not been formally divided into sections or chapters as textbooks are; rather each article is loosely related to the ones near it. It is hoped that if the book loses in orderliness it will gain in freshness, and that perhaps the specialist physicist, the earnest sociologist, and the young reader may thereby be lured into browsing over topics they might otherwise ignore.

Dr Dorothy Fisher and the editorial staff at The Institute of Physics in Bristol have been both stimulating and patient. Mr Hall and Dr Emerson in particular have guided production and accumulated the copyright permissions, which for a manuscript of about 150 separate items is no light undertaking. The designer, Bernard Crossland, evolved a design of sufficiently great adaptability, at first a seemingly impossible task. To all these people, and to the librarians who have helped us trace obscure material and those authors who have contributed special articles, Professor Weber and I are deeply grateful.

ERIC MENDOZA

# When does jam become marmalade?

H  B  G  CASIMIR

A speech delivered by the author at a dinner of the Institute of Electrical Engineers, 1965.

I should like to speak to you for a moment about the problem of two cultures so eloquently formulated by C P Snow and more specifically about jam and marmalade.

A few years ago I visited Istanbul. I was staying at the Hilton Hotel, one of those places that are now all over the world setting a rather high standard of what I consider a rather inferior way of living. One morning at breakfast a very British lady was sitting at a table next to mine. 'Waiter, can I have some marmalade?' she asked peremptorily. A smiling Turkish waiter appeared with a huge tray heavily loaded with some thirty or forty kinds of fruit preserve. The lady looked at them, her face expressing both unbelief and disgust and then said contemptuously: 'Oh, no, those are jam, not marmalade, we never eat jam for breakfast.' It may strike you as funny that this struck me as funny. The point is that in the Dutch language jam is considered to be a very general genus of which orange marmalade is just one subspecies. The strongest statement a Dutchman could possibly make would be: 'The only jam I take at breakfast is orange marmalade' and that is much less categorical. Now it is a curious fact that what may appear to be an arbitrary linguistic convention has a strong influence on our way of thinking. Ask a Dutchman and he will patiently explain that marmalade is made like any other jam by boiling crushed or cut up fruit with sugar, that its taste is both sweet and sour, that it is viscous and sticky. Ask an Englishman and he will equally patiently explain how a particular taste and texture make marmalade a very different thing.

Perhaps it is the amazing richness of the language which tempts the English to make distinctions where others look for general concepts. Let me give a few examples. There are circumstances when it may be very impolite to call a hound a dog or a pony a horse, and a man may not care for billiards but enjoy an occasional game of snooker. I once read an amusing article—by an Englishman of course—on common American misconceptions about England. There was a passage that went roughly as follows: '(A common misconception is) that our beer is sour, flat and lukewarm. On the contrary our beer is bitter, still and served with the chill off. It is served that way because that is the way to serve it. There exists a stuff called lager so tasteless that it can be cooled without damage and so unsubstantial that a few bubbles make no difference. But we don't drink lager, we drink beer.'

A more serious example. We continentals interpret the word 'Europe' to include the British Isles; the British usually do not. I

I

once saw side by side the French and English versions of a book on birds, one being a verbatim translation of the other. The French book was called 'Les oiseaux Européens,' the English version: 'Birds of Europe and the British Isles.' I hope that this linguistic habit will not lead us to emphasize differences and to forget how much we all have in common in historical and cultural background and in the roots of our languages and civilization.

Now I should like to suggest that the so-called difference between the two cultures is largely a case of jam and marmalade. There exists in Dutch, in German, in the Scandinavian languages, a word Wetenschappen, Wissenschaften, Videnskaber that includes all branches of learning. In English *science* usually refers to the natural sciences only. And true enough: what happened with marmalade happens here. We Dutchmen will emphasize the common elements in all *wetenschappen*: the collecting and systematic arranging of data, the search for general principles and for relations between initially unrelated subjects, the willingness to dedicate one's efforts to the pursuit of objective knowledge and so on. A scholar and a natural scientist are both 'wetenschappelijk' because they accept similar criteria, have in many ways a similar attitude. On the other hand, just as the conventional use of English tends to strengthen the differences in appreciation for jam and marmalade or for beer and lager it also leads to overemphasizing the differences between the two branches of learning. But whereas the lady who refuses to eat any kind of jam at breakfast is only mildly ridiculous, the scholar who says he detests any kind of science is not only ridiculous: his attitude is decidedly harmful. Harmful because it encourages those who are responsible for decisions that may determine the fate of mankind to be intentionally ignorant about the material background against which their decisions should be taken. Harmful also because authors and scholars, while gladly using modern commodities, fail to see the philosophical implications of science and tend to deny scientists and engineers their legitimate place in culture.

But we, scientists and engineers, we know that we have not only created material things and above all we know that we contribute to better relations between nations and peoples. For us it is easy to have understanding of and objective appreciation for the work of others, and from there it is not difficult to arrive also at human understanding and appreciation.

Kipling has said that 'there is neither East nor West, Border nor Breed nor Birth, when two strong men stand face to face, though they come from the end of the earth.' I do not hold with that: I profoundly distrust those strong men. But replace 'two strong men' by 'two competent electrical engineers' and though you slightly mar the rhythm you considerably improve the content.

## In defence of pure research

J J THOMSON

From *J J Thomson and the Cavendish Laboratory in His Day* by G P Thomson (New York: Doubleday) 1965 pp 167-8.

[*The following is from a speech Sir J J Thomson made on behalf of a delegation from the Conjoint Board of Scientific Studies in 1916 to Lord Crewe, then Lord President of the Council.*]

By research in pure science I mean research made without any idea of application to industrial matters but solely with the view of extending our knowledge of the Laws of Nature. I will give just one example of the 'utility' of this kind of research, one that has been brought into great prominence by the War—I mean the use of x-rays in surgery. Now, how was this method discovered? It was not the result of a research in applied science starting to find an improved method of locating bullet wounds. This might have led to improved probes, but we cannot imagine it leading to the discovery of x-rays. No, this method is due to an investigation in pure science, made with the object of discovering what is the nature of Electricity. The experiments which led to this discovery seemed to be as remote from 'humanistic interest'—to use a much misappropriated word—as anything that could well be imagined. The apparatus consisted of glass vessels from which the last drops of air had been sucked, and which emitted a weird greenish light when stimulated by formidable looking instruments called induction coils. Near by, perhaps, were great coils of wire and iron built up into electro-magnets. I know well the impression it made on the average spectator, for I have been occupied in experiments of this kind nearly all my life, notwithstanding the advice, given in perfect good faith, by non-scientific visitors to the laboratory, to put that aside and spend my time on something useful.

[*G P Thomson says that he has heard his father use another example, that if Government laboratories had been operating in the Stone Age we should have wonderful stone axes but no-one would have discovered metals!*]

3

# Keeping up with science

LÁSZLÓ FELEKI

Condensed from paper in *Impact of Science on Society* 19, 279 (1969). Published by UNESCO.

With the invention of the steam engine the hell of science broke loose. Since then one admirable discovery has followed the other. Today no human brain is capable of comprehending the whole of science. Today there are part-sciences with part-scientists. Man has hopelessly surpassed himself. He can be proud of this, but he is no longer able to keep track of his own achievements.

Our life has become so mechanized and electronified that one needs some kind of an elixir to make it bearable at all. And what is this elixir if not humour? It is decisive for the present and future of mankind whether humour and science can keep in step, whether there will be time to tell a joke during a journey between two planets, and whether the savant will feel like laughing while he is making efforts to use space for peaceful purposes.

The question 'what is humour?' is one of extraordinary importance; we need to clarify the basic concepts to begin with. To laugh at a joke without analysing it is work half done.

The term 'humour' itself means fluid or moisture, indicating that already the ancient Greeks must have known both moisture and humour. Humour as a fluid probably served to dilute the hard facts of life making it possible to swallow and digest them. Humour is, of course, palatable even without moisture; in such cases we are dealing with dry humour.

One of the characteristics of humour is that it eludes definition. Some partial truths about humour are nevertheless recognizable and I will now cite them.

For instance, it is evident that humour is difficult to write and therefore is certainly not 'light' literature.

Parody is a humorous genre of literature. A really good parody or take-off is better than the original.

The basis of acid humour is ulcers. Many humorists have ulcers.

Truth is often humorous simply because it is so unusual that it makes people laugh.

The greatest blessing of humour is that it relaxes tension. It is really indispensable in situations when there is nothing left but a big laugh (*cf* current history).

Just as the disease of the horse can be demonstrated on a single mare at a veterinary school, by the same token a single joke is suitable for the analysis of all the tenets of the science of humorology. I myself discovered this important fact by mere chance. I told a joke to an acquaintance, who is, by the way, an officer of the Humorology Department of the Hungarian Academy of Sciences . . .

'Well, do you know the one,' I began, 'in which two geologists converse in a cafe? One of them says: "Yes, unfortunately fifteen billion years from now the Sun will cool, and then all life on Earth will perish." A card-player nearby has been half listening to the joke, and turns in terror to the geologist: "What did you say? In how many years will the Sun cool?" "Fifteen billion years," the scientist replies. The card-player lets out a sigh of relief: "Oh, I was afraid you said fifteen million!" '

When I completed the joke to the best of my histrionic ability, I expected the professor to laugh, for it is a delightful little joke, I think. However, instead of the expected smile or laugh my man seemed to be in a brown study—rock-bottom humiliation for a teller of jokes. I was just beginning to think that the professor had not understood the joke, which would not have been too surprising, after all, as humorology was his profession. My supposition, however, proved to be erroneous. A few seconds later the professor gave an appreciative nod.

'The joke is good,' he said. 'If we accept Aristotle's definition according to which the comic, the ridiculous is some fault, deficiency or ugliness which nonetheless causes no pain or trouble, we will find the joke just heard meets these criteria. The cooling of the Sun is certainly a deficiency, or more accurately heat deficiency, although it is not ugliness, for even a chill celestial object can be a very pleasing sight as there are several examples in the universe to demonstrate.

'And, then, what about Hobbes's hypothesis? In his treatise on the causes of laughter Hobbes pointed out that laughter is the feeling of pride as, seeing the weakness of others, we experience our own intellectual superiority.

'The joke also satisfies the contrast theory. For, according to Kant, contrast is the essence of the comic. And in fact it would be difficult to imagine a sharper contrast than that existing between the ephemeral life of man and cosmic time.

'In Schopenhauer's terms, this can also be taken as the disharmony of a concept with some realistic object with which it is associated. Indeed, the card-player who sighs with relief at the idea that he can calmly continue his card-playing until the 14 millionth year of his life, for it will remain warm enough, entertains a most unrealistic thought within the context of a most realistic idea that men like to live as long as possible and dislike to be cold.

'Nor is Bergson's theory of automatism left out of account,

because the protagonist is jolted out of the mechanically induced natural time sense that measures human life.

'To sum it up, I repeat that the joke is funny. Hence I am fully justified in laughing at it.'

And at this moment my friend started to laugh so hard that his tears flowed and he held his sides.

It was easy to laugh in the past at the modest jokes which involved the Little Idiot, the two travelling salesmen, someone's mother-in-law, the drunk, or the Scotsman. Only a small surprise element had to be provided for the listener. A proper appreciation of *scientific* humour requires the proper scientific qualifications. The vital need of future generations is for a scientific education so they can have the incomparable surcease of humour in order to endure the state of perfection to which man and life will have been reduced by the progress of science.

Just consider what degree of culture and education is required to understand the joke which is said to have practically drawn tears of laughter from Einstein and Oppenheimer. One photon asks the other photon weaving about in space: 'Can't you move straight? You must be drunk again!' The other photon protests vehemently: 'What do you expect? Can't you see that I am getting soaked in a gravitational field?' Yes, this is coming, this is what we have to get prepared for.

## Sir Francis Simon, low temperature physicist

N KURTI

From N Kurti, 'Franz Eugen Simon,' *Bio-graphical Memoirs of Fellows of the Royal Society* 4, 225 (1958).

Simon was well known for his ability to clarify issues or to solve controversies by a single apt remark. At committee meetings his interventions were usually brief and to the point. On one occasion committee members were asked by the chairman, who was also in charge of the project, to agree that a certain machine be run at a power which was ten per cent lower than the design value. Simon objected, arguing that 'design value' should mean what it said. Thereupon the chairman remarked: 'Professor Simon, don't you see that we are not talking about science, but about engineering which is an art.' Simon was persistent: 'What would happen if the machine were run at full power?' 'It might get too hot.' 'But, Mr Chairman', came Simon's rejoinder, 'Can't artists use thermometers?'

6

# Cuts by the score

ANON

*NPL News* **236**, 17 (1969).

[*Organization and Method research is carried out to improve the efficiency of working of groups of people. The following are extracts from a report by O & M after a visit to the Royal Festival Hall.*]

For considerable periods the four oboe players had nothing to do. Their numbers should be reduced, and the work spread more evenly over the whole of the concert, thus eliminating peaks of activity.

All the twelve first violins were playing identical notes. This seems unnecessary multiplication. The staff of this section should be drastically cut; if a large volume of sound is required, it could be obtained by means of electronic amplifiers.

Much effort was absorbed in the playing of demisemiquavers. This seems an excessive refinement. It is recommended that all notes should be rounded up to the nearest semiquaver. If this were done it would be possible to use trainees and lower grade operatives more extensively.

There seems to be too much repetition of some musical passages. Scores should be drastically pruned. No useful purpose is served by repeating on the horns a passage which has already been handled by the strings. It is estimated that if all redundant passages were eliminated the whole concert time of two hours could be reduced to twenty minutes, and there would be no need for an interval.

The Conductor agrees generally with these recommendations, but expresses the opinion that there might be some falling-off in box-office receipts. In that unlikely event it should be possible to close sections of the auditorium entirely, with a consequential saving of overhead expense—lighting, attendants, etc.

If the worst came to the worst, the whole thing could be abandoned and the public could go to the Albert Hall instead.

## The theorist

From *Physicists continue to laugh*, MIR Publishing House, Moscow 1968. Translated from the Russian by Mrs Lorraine T Kapitanoff.

When a theoretical physicist is asked, let us say, to calculate the stability of an ordinary four-legged table he rapidly enough arrives at preliminary results which pertain to a one-legged table or a table with an infinite number of legs. He will spend the rest of his life unsuccessfully solving the ordinary problem of the table with an arbitrary, finite, number of legs.

# The theory of practical joking— its relevance to physics

R V JONES

Part of a lecture published in *Bulletin* of the Institute of Physics, June 1957, p 193.

At first sight there may seem little relation between physics and practical joking. Indeed, I might never have observed their connection but for an incidental study of the life of James Clerk Maxwell. Two things, among many others, struck me. The first was the growth of his sense of fun from the primitive joke of the boy of six tripping up the maid with the tea tray to the refined, almost theoretical, jokes of his later life. The second was his mastery of analogy in physical thinking: already, at the age of twenty-four he had written a part playful, part serious essay on the theory of analogy which showed two of the main features of his mind. On the lighter side, he pointed out the relation between an analogy and a pun: in the former one truth lies under two expressions, and in the latter two truths lie under one expression. Hence from the theory of analogy one can by reciprocation deduce the theory of puns. To the more serious side of Maxwell's understanding of analogy I shall return later, but all this set me thinking about the possible connection between the theory of practical joking and physics. One factor which encouraged me was the high incidence of mischievous humour among physicists. Even Newton, it is recorded, caused trouble in his Lincolnshire village as a boy by flying at night a kite carrying a small lantern; and in this century the spritely skill of the late Professor R W Wood and Professor G Gamow is already legendary. While I hope to illustrate this paper with examples, I propose first to analyse (if this is not altogether too brutal a process) the essentials of a joke.

## INCONGRUITIES

The crux of the simplest form of joke seems to be the production of an incongruity in the normal order of events. We hear the story, for example, of Maxwell showing Kelvin some optical experiment, and inviting Kelvin to look through the eyepiece. Kelvin was surprised to find that, while the phenomenon described by Maxwell was undoubtedly there, so was a little human figure, the incongruity, dancing about. Kelvin could not help asking 'Maxwell— but what is the little man there for?' 'Have another look, Thomson,' said Maxwell, 'and you should see.' Kelvin had another look, but was no wiser. 'Tell me, Maxwell,' he said impatiently, 'What *is* he there for?' . . . 'Just for *fun*, Thomson,' replied Maxwell. When we consider a simple incongruity of this type, we can see why this form of humour is sometimes described as 'nonsense'; for 'sense' implies the normal order of things, and in this order an incongruity makes 'nonsense.' A simple incongruity in the literature of physics is R W Wood's recording of the fact that he cleaned out an optical instrument by pushing his cat through it.

Even a change of dimension is sufficient to cause an incongruity. Lord Cherwell has a story of a scientist at Farnborough in World War I, who was so dismayed by the delays in ordering commercial equipment that when he wanted a dark-room lamp he made a pencil sketch of one, to be made up by the workshop. It availed him little, however, because a proper engineer's drawing had by regulation to be made in triplicate before the workshop would start. Weeks elapsed, and finally after a knock on his door two workmen wheeled in the largest dark-room lamp ever constructed. In making the workshop drawing the draughtsman had left out one dash, with the result that intended inches became actual feet. One of the classic incongruities of this type is that due to Benjamin Franklin in a letter to the Editor of a London newspaper in 1765, chaffing the English on their ignorance of America: 'The grand leap of the Whale up the Falls of Niagara is esteemed, by all who have seen it, as one of the finest spectacles in Nature!'

A variation on the simple incongruity in humour is to produce a congruity where incongruity is normally expected. One does not expect, for example, any congruity about the names of joint authors of scientific papers. It was therefore rather a surprise to find a genuine paper by Alpher, Bethe and Gamow, dated April 1 in *The Physical Review* for 1948.

A further variation of humour is produced when a false incongruity is expected by the victim, and an incongruity then genuinely occurs which he promptly discounts. The late Sir Francis Simon had this happen to him when he was head of a laboratory in Germany. One night his research students were working with liquid hydrogen, and there was an explosion which damaged the laboratory some time after midnight. One of the research students telephoned the professor to inform him of the damage. All he could get from Sir Francis was an amiable 'All right, I know what day it is!' It was the morning of April 1.

HOAXES

Simple incongruities, direct or inverted, can be humorous enough, but the more advanced jokes usually involve a period of preparation and induction, sometimes elaborate, before the incongruity becomes apparent. They are called hoaxes. Maxwell's jokes were often simple in their preparation; he is credited with having engineered the advertisement of his Inaugural Lecture at Cambridge (which is still very worth reading) in such a manner that only his undergraduate students heard of it, and he gave it to them alone. The senior members of the University merely saw that the new professor would deliver his first lecture on a particular day, and they attended in force. This lecture, however, was the first of his undergraduate course, and his delighted students enjoyed the experience of seeing Maxwell gravely expounding, though with a betraying twinkle in his eye, the difference between the Fahrenheit and Centigrade scales to men like Adams, Cayley, and Stokes.

With some hoaxes the period of induction of the victim may be extended. In this type, which is probably the most interesting philosophically, the object is to build up in the victim's mind a false world-picture which is temporarily consistent by any tests that he can apply to it, so that he ultimately takes action on it with confidence. The falseness of the picture is then starkly revealed by the incongruity which his action precipitates. It has not proved difficult, for example, to persuade a Doctor of Philosophy to lower his telephone carefully into a bucket of water in the belief that he was cooperating with the engineer in the telephone exchange in finding a leak to earth. The prior induction consisted of building up in his mind a picture of something being wrong with his telephone by such tactics as repeatedly ringing the bell and then ringing off as he answered.

As a further example, we may recall one of the works of a German physicist, Dr Carl Bosch, who about 1934 was working as a research student in a laboratory which overlooked a block of flats. His studies revealed that one of the flats was occupied by a newspaper correspondent, and so he telephoned this victim, pretending to be his own professor. The 'professor' announced that he had just perfected a television device which could enable the user to see the speaker at the other end. The newspaper man was incredulous, but the 'professor' offered to give a demonstration; all the pressman had to do was to strike some attitude, and the voice on the telephone would tell him what he was doing. The telephone was, of course, in direct view of the laboratory, and so all the antics of the pressman were faithfully described. The result was an effusive article in the next day's paper and, subsequently, a bewildered conversation between the true professor and the pressman.

The induction of the victim can take many forms. One of the favourite ways is an acclimatization by slow change. R W Wood is said to have spent some time in a flat in Paris where he discovered that the lady in the flat below kept a tortoise in a window pen. Wood fashioned a collecting device from a broom-handle, and bought a supply of tortoises of dispersed sizes. While the lady was out shopping, Wood replaced her tortoise by one slightly larger. He repeated this operation each day until the growth of the tortoise became so obvious to its owner that she consulted Wood who, having first played a subsidiary joke by sending her to consult a Professor at the Sorbonne whom he considered to be devoid of humour, advised her to write the press. When the tortoise had grown to such a size that several pressmen were taking a daily interest, Wood then reversed the process, and in a week or so the tortoise mysteriously contracted to its original dimensions.

## HOAXES IN WAR

Induced incongruities have a high place in warfare, where if the enemy can be induced to take incorrect action the war may be advantageously affected. A stratagem in which some of my war-time colleagues were involved is now well known as 'The man who never was.' These same colleagues also worked with me in some technical deceptions, of which one was the persuasion of the Germans in 1943 that our successes against the U-boats were due not to centimetric radar but to a fictitious infrared detector. We gained some valuable months while the Germans invented a beautiful anti-infrared paint and failed to find the true causes of their losses. The paint, incidentally, was a Christiansen filter of powdered glass in a transparent matrix over a black base. The filter 'peaked' in the near infrared, so that incident radiation in this region went through and was absorbed in the underlying black. Visible light was scattered back by the filter, which thus gave a light grey appearance to the eye, but was black to the near infrared. This simulated admirably the reflecting power of water, and thus camouflaged the U-boat. It was afterwards reported that the inventor of the paint was Dr Carl Bosch.

Before I turn to the more serious side of this lecture there is one further story from Physics in which the exact classification of the incongruity can be left as a problem to be worked out at leisure. It concerns Lord Kelvin's lectures at Glasgow, where he used to fire a bullet at a ballistic pendulum; as an undergraduate at Oxford I had heard a story of how Kelvin missed on one occasion, with the result that the bullet went through a wall and smashed the blackboard of the lecturer next door. Kelvin rushed into the next room in some alarm to find the lecturer unscathed, and the class shouting 'Missed him—try again, Bill.' This experiment has now produced a further incident, and to avoid any doubt I wrote to Professor Dee for his own account of what happened. This is what he says:

'In the Quincentenary Celebrations here I had to lecture on the history of the Department. Of course Kelvin figured strongly in this. One of Kelvin's traditional experiments was to fire a rifle bullet at a very large ballistic pendulum. All his students regarded this as the highlight of the course. He was reputed to have the gun charged with a big dose of powder—the barrel is about half an inch internal diameter. I decided this experiment must be repeated but there was great alarm here that the barrel would burst and annihilate the front row (Principal and Senate). So I decided to use a modern rifle. I also decided to make it a double purpose experiment by using Kelvin's invention of the optical lever to display the pendulum swing to a large audience. On the night all went off well.

'The next day I repeated the whole lecture to the ordinary class. Mr Atkinson was the normal lecturer to this class and he had noticed that in referring to the dual purpose of the demonstration I used the phrase ". . . fitted a mirror to the pendulum so that I may kill two birds with one stone." After the explosion to my surprise a pigeon fell with a bloody splash on to a large white paper on the bench—our lecture room is very high. I tried to resolve the situation by saying "Well although Mr. Atkinson isn't lecturing to you today he appears to be behind the scenes somewhere. But he does seem to have failed to notice that I said *two* birds with one stone!" Immediately a second pigeon splashed on the bench! Whether this was due to a slip up in Atkinson's mechanical arrangements or to his brilliant anticipation of how I would react I don't really know but I always give him the credit of the second explanation.

'Anyway the students loved it but I wonder how many would remember about the optical lever?'

## TECHNICAL SPOOF IN WAR

I want to turn now to technical deception in war, as exemplified by our attempts to mislead the German night defences in their appreciation of our raiding intentions. The method here is that of the induced incongruity; by a false presentation of evidence we wish the enemy controller to build up an incorrect but self-consistent world-picture, thus causing him to generate the incongruity of directing his nightfighters to some place where our bombers are not. I originally developed this 'Theory of Spoof' in a wartime report; the salient points, which have some interest in physical theory, are the following. As with all hoaxes the first thing is to put oneself in the victim's place (indeed, a good hoax requires a sympathetic nature), to see what evidence he has with which to construct and test his world-picture. In night aerial warfare in 1939-45, this evidence was mainly the presence of deflections in the trace of the cathode ray observing tube. Therefore any device which would give rise to such deflections could provide an element of Spoof. One such device was a jammer which would cause fluctuating deflections all the time, thus concealing the true deflections due to the echo from an aircraft. This, like a smoke screen, would render the enemy unaware that you are where you are. A more positive technique is to provide a false echo, and if possible to suppress the genuine one, thus giving him the impression that you are where you are not. The easiest way of providing a false echo is to drop packets of thin metal strips, cut to resonate to the enemy's radar transmissions. This is, of course, what we did in 1943. There is little time to tell now of the fortunes of this technique, but the packets were extremely successful, and they changed the character of air warfare at night. At first, the German controllers confused the individual packets with aircraft; I can still remember the frustrated tones of one controller repeatedly ordering a packet to waggle its wings as a means of identification. Soon, however, the Germans gave up the attempt to make detailed interceptions, and tried to get a swarm of fighters into our bomber streams. We then used many tinfoil packets dropped by a few aircraft to provide the appearance of spoof raids, which lured the nightfighters off the track of our main raids.

As the war went on the Germans gradually found ways of distinguishing between echoes from metal foil packets and those from aircraft. The packets, for example, resonated to one particular frequency, and therefore they had a relatively poor response to another frequency. If two radar stations watched on widely separate frequencies, a genuine aircraft echo would be present on both, whereas the foil echo would appear only on one. The foil could, of course, be cut to different lengths, but as the number of frequencies was increased, the amount of foil needed was greater. Moreover there was a pronounced Doppler effect on the echo from an aircraft, with its high speed, but little effect on the echoes from the foil drifting with the wind. Thus, against an omniscient controller, we have to make the decoy echoes move with the speed of aircraft, and reflect different frequencies in the same way. This is easiest done by making a glider of the same size as the bomber. Then if we allow the enemy controller to use sound and infrared detectors and other aids, we find that the only decoy which can mislead him into thinking that there is a British bomber flying through his defences is another British bomber flying through his defences.

Another example is one that I encountered earlier in what has been called 'The Battle of the Beams' in 1940. Here the problem was to upset the navigation of the German night bombers, when they were flying along radio beams to their targets. The signals received by the pilots telling them to steer right or left were counterfeited in this country, and sometimes resulted in their flying on curvilinear courses. However, had the pilots had un-limited time of observation they could have detected that there was something wrong, even if we had exactly synchronized our transmitters with those of the Germans. The bombers were in general flying away from their own transmitters and towards ours, and so they would have received a Doppler beat from which they could have deduced that a second transmitter was active. If one allows the possibility of various simple tests, which fortun-ately would take too long in actual warfare, one arrives at the conclusion that the only place for a second transmitter which will simulate the original exactly is coincident with the original and the counterfeit thus defeats its purpose.

A physicist had a horseshoe hanging on the door of his laboratory. His colleagues were surprised and asked whether he believed that it would bring luck to his experiments. He answered: 'No, I don't believe in superstitions. But I have been told that it works even if you don't believe in it.'

*[Told by I B Cohen, the Harvard historian of physics, to S A Goudsmit who told it to Bohr, whose favourite story it became.]*

# New University—1229

Condensed from
*University Records
of the Middle Ages,*
by Lynn Thorn-
dike (Columbia
University Press,
1944).

[*In 1229 a new University was founded at Toulouse, and this advertisement was issued (in Latin of course). Prominent among the works of
Aristotle forbidden at Paris but studied at Toulouse was his 'Physics'.*]

The Lord Cardinal and Legate in the Realm of France, leader and
protector and author after God and the Pope of so arduous a
beginning . . . decreed that all studying at Toulouse, both
masters and disciples, should obtain plenary indulgence of all their
sins.

Further, that ye may not bring hoes to sterile and uncultivated
fields, the professors at Toulouse have cleared away for you the
weeds of the rude populace and the thorns of sharp sterility and
other obstacles. For here theologians in pulpits inform their disciples and the people at the crossroads, logicians train beginners
in the arts of Aristotle, grammarians fashion the tongues of the
stammering on analogy, organists smooth the popular ears with
the sweet-throated organ, decretists extol Justinian, and physi-
cians teach Galen. Those who wish to scrutinize the bosom of
nature to the inmost can hear here the books of Aristotle which
were forbidden at Paris.

What then will you lack? Scholastic liberty? By no means, since
tied to no-one's apron strings you will enjoy your own liberty. Or
do you fear the malice of the raging mob or the tyranny of an
injurious prince? Fear not . . .

As for fees, what has already been said and the fact that there is
no fear of a failure of crops should reassure you. The courtesy of
the people should not be passed over. So if you wish to marvel at
more good things than we have mentioned, leave home behind,
strap your knapsack on your back . . .

## The Smithsonian Institution

I would like to know of what this Institution consists. I would like
the gentleman from New York or the gentleman from Vermont to
tell us how many of his constituents ever saw this Institution or
ever will see it or ever want to see it? It is enough to make any
man or woman sick to visit that Institution. No one can expect to
get any benefit from it.

Lewis Selye, House of Representatives, 1868.

# Atmospheric extravaganza

JOHN HERAPATH

Condensed from
Railway Journal 7,
2769–71; 8, 17–19,
83, 115, 116.

[*John Herapath enters the history of science as one of the progenitors of the kinetic theory of gases. His main interest was in fact the development of railways; in 1838 he founded and for many years edited the* Railway Journal. *Early issues are odd assortments of financial analyses, discussions of mechanisms and fundamental physics—exposures, diatribes and mathematics make up the rumbustious mixture. The skit which follows is directed against the Atmospheric Railway, promoted by Brunel.*

*In this system, the train ran on rails in the usual way but the locomotive power was supplied by vacuum. A pipe (27 inches in diameter on the London and Croydon) was laid between the rails, with a slit at the top all along its length. Inside the pipe was a piston, connected to the train by a rod running through the slit. The slit was closed by a 'valve', a leather strip, raised automatically as the train went by so that the rod could pass along. The pipe was evacuated at one end by a large pump driven by a steam engine and the train was driven along. The 'atmospheric' was particularly useful for working steep gradients; lines laid in the West of England and in Ireland functioned for several years, one in France till 1860. Two problems defeated the system: the puckering of the leather strip under atmospheric pressure and the impossibility of finding vacuum grease which neither melted nor hardened with the weather (and was not eaten by rats).*]

The year is 1845. Samuda–later a distinguished shipbuilder and naval architect—and Wilkinson are two Directors of the Company. They are taking a party of Shareholders for a demonstration ride.

*The party is arrived, and Samuda goes into the engine-house.*

SAMUDA: Well, have you a good vacuum?

FIRST MAN: No, Sir, we can't get a good one, nor scarcely any at all, and yet we have been pumping for hours.

SAMUDA: How is that?

FIRST MAN: Why, you know, Sir, it is one of our common occurrences. I have pulled the governors off, and driven the engines to 40 or 50, instead of 18 or 20 strokes a minute. I have actually been afraid the engines would fly to pieces, and the house come down upon us, and here we are as we were three hours ago, and getting worse rather than better.

SAMUDA: Confound it, how unlucky. We must do something today.

SECOND MAN: Sir, the Sun has melted the grease and it has all run into the tubes and choked them up. The valve, too, has puckered up and all of us together can't keep it down, though I have menon, the line as thick as blackberries.

SAMUDA: The Sun, man! How dare the Sun melt my grease? I tell you, my

16

grease (composition, I mean) will neither soften with heat nor harden with cold.

SECOND MAN: That may be so among the soft 'uns in the House of Commons; but here we find it different. The grease not merely melts but actually runs away with a little Sun, as it has to-day.

WILKINSON (*outside*): Come, Samuda, the gentlemen are impatient for the ride.

SAMUDA (*to the Men*): Put on all your steam; work away as hard as you can. Spare nothing to give us a good high speed. Now or never we must make a splash today.

*Goes out. The party is seated. They start, and the train crawls along at the rate of seven or eight miles an hour.*

SHAREHOLDERS: Mr. Samuda, I thought we were to go at a high speed. You see there goes the Dover train flying past us like the wind. Why, we can't be going above six or seven miles an hour.

SAMUDA: It ill becomes me, in my humble situation, to controvert the opinions of you, my illustrious masters; but I assure you we are going at least 20 miles an hour.

SHAREHOLDERS: Why, we have been six or eight minutes going one mile.

SAMUDA'S FIRST SATELLITE: No, gentlemen, pardon me; I have taken the time very carefully, and I find we have been just 2 minutes, 59 seconds, a half, a quarter, and 23 hundredths of a second coming the mile; which being reduced, first to decimal, and then to vulgar fractions, and worked by a peculiar arithmetic, the invention, I believe (*bowing*), of that great man, Mr Samuda, comes out 25 miles, a yard, an inch, and a barley-corn per hour.

SECOND SATELLITE: That, gentlemen, is very near the truth. My time is just half a hundredth of a second more, which, by calculation, gives exactly seven-eighths of a barley-corn per hour less velocity.

SAMUDA (*to the Shareholders*): My honoured masters, you hear what these two very credible gentlemen say. Their close agreement and great accuracy must prove to you that they are right. Something must have affected your watches. I have found it so in more cases than one. Then as to the Dover train passing us, that was, I assure you, an optical illusion; possibly a reflection of ourselves from the concave, transparent, cerulean, ether of the sky.

SHAREHOLDERS: But still we should like to go a little faster; perhaps we shall, returning.

SAMUDA: Oh! yes, certainly, a 100 miles an hour, if you desire it. When we get the Portsmouth line, we will show you what we can do. We shall travel at such a rate as to do away with the electric telegraph altogether. We shall become that ourselves, and expect to

derive a large income from that source alone. Indeed, I may here
just tell you—but I don't wish it to be published, and above all, I
don't want it to get to that abominable *Herapath's Journal*,—that
my excellent friend, Brunel, has a most magnificent project to
bring before the next Parliament, to be worked by Atmospheric
power. He intends to propose a railway from here to the East
Indies.

SHAREHOLDERS: But how will he cross the English Channel and the
Mediterranean?

SAMUDA: Oh, they are mere trifles. He will build bridges over them. That
too will be on the Atmospheric principle. Instead of piers the
bridges will be supported at various points by balloons, the gas
being drawn from a pit spontaneously generating it near New-
castle, which may be had for merely the expense of laying down
the pipes. I may as well here just add, that besides the profits,
large tiger preserves in Bengal, mentioned by Dickens, and considerable tracts on the tops of the Himalayan mountains, when
ascended, are to be given as bonuses. I would recommend you,
gentlemen, not to lose this splendid opportunity of making your
fortunes.

*The shareholders are now arrived at Croydon. They examine the
premises and the machinery while Samuda talks to Wilkinson.*

WILKINSON: What a miserable velocity we got up. A good horse would
have walked quite as fast as we came.

SAMUDA: But did I not amuse them nicely about the Anglo–East-Indian
Railway? It was a capital thought, was it not?

WILKINSON: Capital indeed; but I had something to do to keep my
countenance.

*First bell rings for the return, and all hie to the carriages.*

SAMUDA (*to Wilkinson*): Just cast your eye upon those fellows we came
up with. See what part of the train they get into, and we will go
to another, for I don't want to come into contact with them again.
I don't like their questions.

WILKINSON: 'Fore Gad, here they come straight to us.

SAMUDA: Confound them, I wish them a hundred miles off.

SHAREHOLDERS: Well, Mr. Samuda, we are glad we have met with you;
we want to talk to you about the tiger preserves and the bonuses
on the tops of the Himalayan mountains.

SAMUDA: Hush! gentlemen, hush! If it should get abroad, you will be
done out of all these fine things, clean done. You had better, I
think, take your seats, or else the best will be gone.

SHAREHOLDERS: Never mind that. We like your company better than all the seats, and if we are with you we shan't fare badly.

SAMUDA (*aside to Wilkinson*): A plague on them; what shall I do?

*Second bell rings.*

SAMUDA (*aside*): Ill-luck betide them. They have evil designs, I fear.

*They get into a carriage just opposite a dial, to which Samuda very innocently directs attention, under the plea of finding fault with its place. It is a quarter past three by it. They start, and at first go at a better pace than coming.*

SHAREHOLDERS: Well, Mr. Samuda, I hope we shall get up a higher velocity.

SAMUDA: I hope so too, my honoured masters.

*But the journey is no faster than before. They arrive at the station. Some one, pointing to the clock, remarks they had only been five minutes coming the five miles, as it was a quarter past three when they started, and now it is nearly twenty minutes after. It is re-echoed by a legion of satellites but the Shareholders, who had carefully looked at their watches, declare they have been above forty minutes. Samuda hurries away.*

SHAREHOLDERS (*calling after him*): We shall want shortly to have a little conversation with you about your bridges over the English Channel and Mediterranean. Capital notion that; but how do you keep them steady in a gale, if suspended by balloons?

SAMUDA (*capering about in high glee at the felicitous answer he should give, sings out triumphantly*).

Pray, Sirs, yourselves don't alarm,
Nothing our bridges can harm;
'Tis not tempests nor storms can them mar,
For in Brunel's single head
There's a vast deal more lead
Than would anchor the earth to a star,
Than would anchor the earth to a star.

*Chorus all*
For in Brunel's little head
There's a vast deal more lead
Than would anchor the earth to a star

SHAREHOLDERS (*laughing heartily*): Your answer is indisputable.

SAMUDA: Then you will subscribe to our grand project—the Anglo–East-Indian, Brunellian, Wilkinsonian, Atmospheric Railway?

SHAREHOLDERS: We'll think of it; meanwhile we should like to know something of the bonus lands on the tops of the Himalayan mountains. They are five miles high. Is it not excessively cold there? Are they not covered with eternal snow?

SAMUDA: I assure you, you are in error. It is always beautifully fine there, and so warm that you would not complain after you had been there a little while (*Aside*) Anyone would be frozen to death in a few hours.

SHAREHOLDERS: Why, Laplace, Mr. Herapath, and philosophers generally say the heat decreases rapidly as we ascend in the atmosphere; and Mr. Herapath has written that it should be 32° below freezing at the top of the mountains.

SAMUDA: Pooh! pooh! Newton, Laplace, and Herapath, know nothing about it. Use your own sense, my masters, and you will see they are all wrong. Five miles high is above the clouds; how can there be any snow there? Besides, is it not five miles nearer the Sun? Of course it is; and of course it is so much warmer; and the Sun always shines there, and that will make it still warmer.

*Brunel walks past, and episode degenerates into a slanging match.*

From *The Space Child's Mother Goose,* verse by Frederick Winsor, illustration by Marion Parry (New York: Simon and Schuster) 1958.

*Little Miss Muffet*
*Sits on her tuffet*
*In a nonchalant sort of way.*
*With her force field around her*
*The spider, the bounder,*
*Is not in the picture today.*

# The Academy

JONATHAN SWIFT

From *Gulliver's Travels* by Jonathan Swift, Part III '*A Voyage to Laputa*' Chapter 5 (1727).

This Academy (at Lagado) is not an entire single Building, but a Continuation of several Houses on both Sides of a Street; which growing waste, was purchased and applyed to that Use.

I was received very kindly by the Warden, and went for many Days to the Academy. Every Room hath in it one or more Projectors; and I believe I could not be in fewer than five Hundred Rooms.

The first Man I saw was of a meagre Aspect, with sooty Hands and Face, his Hair and Beard long, ragged and singed in several Places. His Clothes, Shirt, and Skin were all of the same Colour. He had been Eight Years upon a Project for extracting Sun-Beams out of Cucumbers, which were to be put into Vials hermetically sealed, and let out to warm the Air in raw inclement Summers. He told me, he did not doubt in Eight Years more, that he should be able to supply the Governor's Gardens with Sun-shine at a reasonable Rate; but he complained that his Stock was low, and intreated me to give him something as an Encouragement to Ingenuity, especially since this had been a very dear Season for Cucumbers. I made him a small Present, for my Lord had furnished me with Money on purpose, because he knew their Practice of begging from all who go to see them.

I saw another at work to calcine Ice into Gunpowder; who likewise shewed me a Treatise he had written concerning the Malleability of Fire, which he intended to publish.

There was a most ingenious Architect who had contrived a new Method for building Houses, by beginning at the Roof, and working downwards to the Foundation; which he justified to me by the like Practice of those two prudent Insects the Bee and the Spider.

In another Apartment I was highly pleased with a Projector, who had found a Device of plowing the Ground with Hogs, to save the Charges of Plows, Cattle, and Labour. The Method is this: In an Acre of Ground you bury at six Inches Distance, and eight deep, a Quantity of Acorns, Dates, Chesnuts, and other Masts or Vegetables whereof these Animals are fondest; then you drive six Hundred or more of them into the Field, where in a few Days they will root up the whole Ground in search of their Food, and make it fit for sowing, at the same time manuring it with their Dung. It is true, upon Experiment they found the Charge and Trouble very great, and they had little or no Crop. However, it is not doubted that this Invention may be capable of great Improvement.

I had hitherto seen only one Side of the Academy, the other being appropriated to the Advancers of speculative Learning.

Some were condensing Air into a dry tangible Substance, by extracting the Nitre, and letting the acqueous or fluid Particles percolate: Others softening Marble for Pillows and Pin-cushions. Another was, by a certain Composition of Gums, Minerals, and Vegetables outwardly applied, to prevent the Growth of Wool upon two young Lambs; and he hoped in a reasonable Time to propagate the Breed of naked Sheep all over the Kingdom.

## The triumph of reason

BERT LISTON TAYLOR

*'A Line o'Type or Two,' Chicago Tribune* c. 1920 and reprinted in *Journal of the Optical Society of America*, May 1964.

*Behold the mighty dinosaur*
  *Famous in prehistoric lore,*
*Not only for his weight and length*
  *But for his intellectual strength.*
*You will observe by these remains*
  *The creature had two sets of brains—*
*One in his head (the usual place),*
  *The other at his spinal base.*
*Thus he could reason* a priori
  *As well as* a posteriori.
*No problem bothered him a bit*
  *He made both head and tail of it.*

*So wise was he, so wise and solemn,*
  *Each thought filled just a spinal column.*
*If one brain found the pressure strong*
  *It passed a few ideas along.*
*If something slipped his forward mind*
  *'Twas rescued by the one behind.*
*And if in error he was caught*
  *He had a saving afterthought.*
*As he thought twice before he spoke*
  *He had no judgment to revoke.*
*Thus he could think without congestion*
  *Upon both sides of every question.*
*Oh, gaze upon this model beast*
  *Defunct ten million years at least.*

# American Institute of Useless Research

From *Review of Scientific Instruments* **6**, 208 (1935).

Dear Sir

The following material is humbly submitted by the Committee in Executive Session and it is sincerely hoped that it will meet with your approval for publication:

'There has long been felt in American Physics the need for an efficient governing body to organize the vast quantity of useless research that is being pursued day by day and hour by hour in the many institutions of higher learning in these great United States for which our forefathers fought and bled. With this end in view there has already been formed a branch of the AIUR at one of the aforementioned institutions of higher learning. It is the fervent hope of the founders that this worthy movement will spread its tentacles throughout the land and be an ever present aid to those endeavoring to unscrew the inscrutable.

'Those who attended the inaugural meeting of the AIUR were fortunate in having the opportunity to hear a well-known authority on Banned Spectra, Professor O H Molecule, who gave a talk on *Some Higher Harmonics in the Brass Bands*.

'In addition to holding various meetings, colloquia and seminars the AIUR proposes to recognize by election to fellowship workers in all fields who have contributed some outstanding piece of Useless Research. The AIUR also sponsors the following Journals: *The Refuse of Modern Physics, The Nasty-physical Journal,* and for those who are unable to read English, the *Comptes Fondues,* and the *Makeshift für Physik*.'

Respectfully,

J J COUPLING

ELLEN S COUPLING

*For the Committee.*

American Institute of Useless Research
M I T Branch
Cambridge, Massachusetts,
29 May, 1935

[*Yardley Beers, of the National Bureau of Standards, writes:*]

The early meetings of the society were held in the bottom of an elevator shaft. One of the first activities was the rewiring of the elevator buttons at MIT so that when you pushed the button marked 1 you went to the 4th floor and so on. The Institute had a

song: 'When night krypton and the stars argon/The moon radon
then you xenon . . .' Typical papers were devoted to such topics
as what to do with post holes once you had dug them, the mathematical theory of ballroom dancing, a project to change the
moment of inertia of the earth to keep the Russians under the sun
and fry them to a frazzle, and so on.

---

## Remarks on the quantum theory of the absolute zero of temperature

G BECK, H BETHE and W RIEZLER

Translated from
*Die Naturwissen-
schaften* **2** (1931)
p 38–9.

[*This is a famous spoof paper, accepted by the Editor of* Die Naturwissen-
schaften *in good faith, and published, in 1931. It pokes fun at the mystical
properties claimed by Eddington and others for the number 137.*]

Let us consider a hexagonal crystal lattice. The absolute zero
temperature is characterized by the condition that all degrees of
freedom are frozen. That means all inner movements of the
lattice cease. This of course does not hold for an electron on a
Bohr orbital. According to Eddington, each electron has $1/\alpha$
degrees of freedom, where $\alpha$ is the Sommerfeld fine structure
constant. Beside the electrons, the crystal contains only protons
for which the number of degrees of freedom is the same since,
according to Dirac, the proton can be viewed as a hole in the
electron gas. To obtain absolute zero temperature we therefore
have to remove from the substance $2/\alpha - 1$ degrees of freedom per
neutron. (The crystal as a whole is supposed to be electrically
neutral; 1 neutron = 1 electron + 1 proton. One degree of
freedom remains because of the orbital movement.)

For the absolute zero temperature we therefore obtain

$$T_0 = - (2/\alpha - 1) \text{ degrees.}$$

If we take $T_0 = -273$ we obtain for $1/\alpha$ the value of 137 which
agrees within limits with the number obtained by an entirely
different method. It can be shown easily that this result is independent of the choice of crystal structure.

# A contribution to the mathematical theory of big game hunting

H PÉTARD *Princeton, New Jersey*

From *American Mathematical Monthly* 45 446 (1938).

This little known mathematical discipline has not, of recent years, received in the literature the attention which, in our opinion, it deserves. In the present paper we present some algorithms which, it is hoped, may be of interest to other workers in the field. Neglecting the more obviously trivial methods, we shall confine our attention to those which involve significant applications of ideas familiar to mathematicians and physicists.

The present time is particularly fitting for the preparation of an account of the subject, since recent advances both in pure mathematics and in theoretical physics have made available powerful tools whose very existence was unsuspected by earlier investigators. At the same time, some of the more elegant classical methods acquire new significance in the light of modern discoveries. Like many other branches of knowledge to which mathematical techniques have been applied in recent years, the Mathematical Theory of Big Game Hunting has a singularly happy unifying effect on the most diverse branches of the exact sciences.

For the sake of simplicity of statement, we shall confine our attention to Lions (*Felis leo*) whose habitat is the Sahara Desert. The methods which we shall enumerate will easily be seen to be applicable, with obvious formal modifications, to other carnivores and to other portions of the globe. The paper is divided into three parts, which draw their material respectively from mathematics, theoretical physics, and experimental physics.

The author desires to acknowledge his indebtness to the Trivial Club of St John's College, Cambridge, England; to the MIT chapter of the Society for Useless Research; to the F o P, of Princeton University; and to numerous individual contributors, known and unknown, conscious and unconscious.

## 1. MATHEMATICAL METHODS

**1. The Hilbert, or axiomatic, method.** We place a locked cage at a given point of the desert. We then introduce the following logical system.

**Axiom 1.** *The class of lions in the Sahara Desert is non-void.*

**Axiom 2.** *If there is a lion in the Sahara Desert, there is a lion in the cage.*

**Rule of procedure.** *If p is a theorem, and 'p implies q' is a theorem, then q is a theorem.*

**Theorem 1.** *There is a lion in the cage.*

**2. The method of inversive geometry.** We place a *spherical* cage in the desert, enter it, and lock it. We perform an inversion with respect to the cage. The lion is then in the interior of the cage, and we are outside.

**3. The method of projective geometry.** Without loss of generality, we may regard the Sahara Desert as a plane. Project the plane into a line, and then project the line into an interior point of the cage. The lion is projected into the same point.

**4. The Bolzano-Weierstrass method.** Bisect the desert by a line running N-S. The lion is either in the E portion or in the W portion; let us suppose him to be in the W portion. Bisect this portion by a line running E-W. The lion is either in the N portion or in the S portion; let us suppose him to be in the N portion. We continue this process indefinitely, constructing a sufficiently strong fence about the chosen portion at each step. The diameter of the chosen portions approaches zero, so that the lion is ultimately surrounded by a fence of arbitrarily small perimeter.

**5. The 'Mengentheoretisch' method.** We observe that the desert is a separable space. It therefore contains an enumerable dense set of points, from which can be extracted a sequence having the lion as limit. We then approach the lion stealthily along this sequence, bearing with us suitable equipment.

**6. The Peano method.** Construct, by standard methods, a continuous curve passing through every point of the desert. It has been remarked [1] that it is possible to traverse such a curve in an arbitrarily short time. Armed with a spear, we traverse the curve in a time shorter than that in which a lion can move his own length.

**7. A topological method.** We observe that a lion has at least the connectivity of the torus. We transport the desert into four-space. It is then possible [2] to carry out such a deformation that the lion can be returned to three-space in a knotted condition. He is then helpless.

**8. The Cauchy, or functiontheoretical, method.** We consider an analytic lion-valued function $f(z)$. Let $\zeta$ be the cage. Consider the integral

$$\frac{1}{2\pi i} \int_C \frac{f(z)}{z - \zeta}\, dz,$$

where C is the boundary of the desert; its value is $f(\zeta)$, *i.e.*, a lion in the cage. [3]

**9. The Wiener Tauberian method.** We procure a tame lion, $L_0$ of class $L(-\infty, \infty)$, whose Fourier transform nowhere vanishes, and release it in the desert. $L_0$ then converges to our cage. By Wiener's General Tauberian Theorem, [4] any other lion, $L$ (say), will then converge to the same cage. Alternatively, we can approximate arbitrarily closely to $L$ by translating $L_0$ about the desert. [5]

2. METHODS FROM THEORETICAL PHYSICS

**10. The Dirac method.** We observe that wild lions are, *ipso facto*, not observable in the Sahara Desert. Consequently, if there are any lions in the Sahara, they are tame. The capture of a tame lion may be left as an exercise for the reader.

**11. The Schrödinger method.** At any given moment there is a positive probability that there is a lion in the cage. Sit down and wait.

**12. The method of nuclear physics.** Place a tame lion in the cage, and apply a Majorana exchange operator [6] between it and a wild lion.

As a variant, let us suppose, to fix ideas, that we require a male lion. We place a tame lioness in the cage, and apply a Heisenberg exchange operator [7] which exchanges the spins.

**13. A relativistic method.** We distribute about the desert lion bait containing large portions of the Companion of Sirius. When enough bait has been taken, we project a beam of light across the desert. This will bend right round the lion, who will then become so dizzy that he can be approached with impunity.

3. METHODS FROM EXPERIMENTAL PHYSICS

**14. The thermodynamical method.** We construct a semi-permeable membrane, permeable to everything except lions, and sweep it across the desert.

**15. The atom-splitting method.** We irradiate the desert with slow neutrons. The lion becomes radioactive, and a process of disintegration sets in. When the decay has proceeded sufficiently far, he will become incapable of showing fight.

**16. The magneto-optical method.** We plant a large lenticular bed of catnip (*Nepeta cataria*), whose axis lies along the direction of the horizontal component of the earth's magnetic field, and place a cage at one of its foci. We distribute over the desert large quantities of magnetized spinach (*Spinacia oleracea*), which, as is well known, has a high ferric content. The spinach is eaten by the herbivorous denizens of the desert, which are in turn eaten by lions. The lions are then oriented parallel to the earth's magnetic field, and the resulting beam of lions is focused by the catnip upon the cage.

1 By Hilbert. See E W Hobson, *The Theory of Functions of a Real Variable and the Theory of Fourier's Series* (1927) vol 1, pp 456–457.

2 H Seifert and W Threlfall, *Lehrbuch der Topologie* (1934) pp 2–3.

3 *N.B.* By Picard's Theorem (W F Osgood, *Lehrbuch der Funktionentheorie*, vol 1 (1928) p 178), we can catch every lion with at most one exception.

4 N Wiener, *The Fourier Integral and Certain of its Applications* (1933) pp 73–74.

5 N Wiener, *loc cit*, p 89.

6 See, for example, H A Bethe and R F Bacher, *Reviews o Modern Physics*, **8** (1936) pp 82–229; especially pp 106–107.

7 *Ibid.*

## Fission and Superstition

[*A cautionary verse for parents or children appropriate to the Christmas season.*]

*New Statesman and Nation* (London) Jan 14, 1950.

*This is the Tale of Frederick Wermyss*
*Whose Parents weren't on speaking terms.*
*So when Fred wrote to Santa Claus*
*It was in duplicate because*
*One went to Dad and one to Mum—*
*Both asked for some Plutonium.*
*See the result: Father and Mother—*
*Without Consulting one another—*
*Purchased two Lumps of Largish Size,*
*Intending them as a Surprise,*
*Which met in Frederick's Stocking and*
*Laid level Ten square Miles of Land.*

MORAL
*Learn from this Dismal Tale of Fission*
*Not to mix Science with Superstition.*

HMK

# The uses of fallacy

PAUL V DUNMORE

Originally
published in
*New Zealand
Mathematics
Magazine* 7, 15
(1970). Revised
by the author.

In the last hundred years or so, mathematics has undergone a tremendous growth in size and complexity and subtlety. This growth has given rise to a demand for more flexible methods of proving theorems than the laborious, difficult, pedantic, 'rigorous' methods previously in favour. This demand has been met by what is now a well-developed branch of mathematics known as Generalized Logic. I don't want to develop the theory of Generalized Logic in detail, but I must introduce some necessary terms. In Classical Logic, a Theorem consists of a True Statement for which there exists a Classical Proof. In Generalized Logic, we relax both of these restrictions: a Generalized Theorem consists of a Statement for which there exists a Generalized Proof. I think that the meaning of these terms should be sufficiently clear without the need for elaborate definitions.

The applications of Generalized Proofs will be obvious. Professional authors of text-books use them freely, especially when proving mathematical results in Physics texts. Teachers and lecturers find that the use of Generalized Proofs enables them to make complex ideas readily accessible to students at an elementary level (without the necessity for the tutor to understand them himself). Research workers in a hurry to claim priority for a new result, or who lack the time and inclination to be pedantic, find Generalized Proofs useful in writing papers. In this application, Generalized Proofs have the further advantage that the result is not required to be true, thus eliminating a tiresome (and now superfluous) restriction on the growth of mathematics.

I want now to consider some of the proof techniques which Generalized Logic has made available. I will be concerned mostly with the ways in which these methods can be applied in lecture courses—they require only trivial modifications to be used in text books and research papers.

The *reductio* methods are particularly worthy of note. There are, as everyone knows, two *reductio* methods available: *reductio ad nauseam* and *reductio ad erratum*. Both methods begin in the same way: the mathematician denies the result he is trying to prove, and writes down all the consequences of this denial that he can think of. The methods are most effective if these consequences are written down at random, preferably in odd vacant corners of the blackboard.

Although the methods begin in the same way, their aims are completely different. In *reductio ad nauseam* the lecturer's aim is to get everyone in the class asleep and not taking notes. (The latter

29

is a much stronger condition.) The lecturer then has only to clean the blackboard and announce, 'Thus we arrive at a contradiction, and the result is established'. There is no need to shout this—it is the signal for which everyone's subconscious has been waiting. The entire class will awaken, stretch, and decide to get the last part of the proof from someone else. If everyone had stopped taking notes, therefore, there is no 'someone else', and the result is established.

In *reductio ad erratum* the aim is more subtle. If the working is complicated and pointless enough, an error is bound to occur. The first few such mistakes may well be picked up by an attentive class, but sooner or later one will get through. For a while, this error will lie dormant, buried deep in the working, but eventually it will come to the surface and announce its presence by contradicting something which has gone before. The theorem is then proved.

It should be noted that in *reductio ad erratum* the lecturer need not be aware of this random error or of the use he has made of it. The best practitioners of this method can produce deep and subtle errors within two or three lines and surface them within minutes, all by an instinctive process of which they are never aware. The subconscious artistry displayed by a really virtuoso master to a connoisseur who knows what to look for can be breathtaking.

There is a whole class of methods which can be applied when a lecturer can get from his premisses $P$ to a statement $A$, and from another statement $B$ to the desired conclusion $C$, but he cannot bridge the gap from $A$ to $B$. A number of techniques are available to the aggressive lecturer in this emergency. He can write down $A$, and without any hesitation put 'therefore $B$'. If the theorem is dull enough, it is unlikely that anyone will question the 'therefore'. This is the method of Proof by Omission, and is remarkably easy to get away with (sorry, 'remarkably easy to apply with success').

Alternatively, there is the Proof by Misdirection, where some statement that looks rather like '$A$, therefore $B$' is proved. A good bet is to prove the converse '$B$, therefore $A$': this will always satisfy a first-year class. The Proof by Misdirection has a countably infinite analogue, if the lecturer is not pressed for time, in the method of Proof by Convergent Irrelevancies.

Proof by Definition can sometimes be used: the lecturer defines a set $S$ of whatever entities he is considering for which $B$ is true, and announces that in future he will be concerned only with members of $S$. Even an Honours class will probably take this at face value, without enquiring whether the set $S$ might not be empty.

Proof by Assertion is unanswerable. If some vague waffle about why $B$ is true does not satisfy the class, the lecturer simply says, 'This point should be intuitively obvious. I've explained it as clearly as I can. If you still cannot see it, you will just have to think very carefully about it yourselves, and then you will see how trivial and obvious it is.'

The hallmark of a Proof by Admission of Ignorance is the statement, 'None of the text-books makes this point clear. The result is certainly true, but I don't know why. We shall just have to accept it as it stands.' This otherwise satisfactory method has the potential disadvantage that somebody in the class may know why the result is true (or, worse, know why it is false) and be prepared to say so.

A Proof by Non-Existent Reference will silence all but the most determined troublemaker. 'You will find a proof of this given in Copson on page 445', which is in the middle of the index. An important variant of this technique can be used by lecturers in pairs. Dr Jones assumes a result which Professor Smith will be proving later in the year—but Professor Smith, finding himself short of time, omits that theorem, since the class has already done it with Dr Jones. . . .

Proof by Physical Reasoning provides uniqueness theorems for many difficult systems of differential equations, but it has other important applications besides. The cosine formula for a triangle, for example, can be obtained by considering the equilibrium of a mechanical system. (Physicists then reverse the procedure, obtaining the conditions for equilibrium of the system from the cosine rule rather than from experiment.)

The ultimate and irrefutable standby, of course, is the self-explanatory technique of Proof by Assignment. In a text-book, this can be recognized by the typical expressions 'It can readily be shown that . . .' or 'We leave as a trivial exercise for the reader the proof that . . .'. (The words 'readily' and 'trivial' are an essential part of the technique.)

An obvious and fruitful ploy when confronted with the difficult problem of showing that $B$ follows from $A$ is the Delayed Lemma. 'We assert as a lemma, the proof of which we postpone. . . '. This is by no means idle procrastination: there are two possible dénouements. In the first place, the lemma may actually be proved later on, using the original theorem in the argument. This Proof by Circular Cross-Reference has an obvious inductive generalization to chains of three or more theorems, and some very elegant results arise when this chain of interdependent theorems becomes infinite.

The other possible fate of a Delayed Lemma is the Proof by Infinite Neglect, in which the lecture course terminates before the lemma has been proved. The lemma, and the theorem of which it is a part, will naturally be assumed without comment in future courses.

A very subtle method of proving a theorem is the method of Proof by Osmosis. Here the theorem is never stated, and no hint of its proof is given, but by the end of the course it is tacitly assumed to be known. The theorem floats about in the air during the entire course and the mechanism by which the class absorbs it is the well-known biological phenomenon of osmosis.

A method of proof which is regrettably little used in undergraduate mathematics is the Proof by Aesthetics ('This result is too beautiful to be false'). Physicists will be aware that Dirac uses this method to establish the validity of several of his theories, the evidence for which is otherwise fairly slender. His remark 'It is more important to have beauty in one's equations than to have them fit experiment'[1] has achieved a certain fame.

I want to discuss finally the Proof by Oral Tradition. This method gives rise to the celebrated Folk Theorems, of which Fermat's Last Theorem is an imperfect example. The classical type exists only as a footnote in a text-book, to the effect that it can be proved (see unpublished lecture notes of the late Professor Green) that. . . . Reference to the late Professor Green's lecture notes reveals that he had never actually seen the proof, but had been assured of its validity in a personal communication, since destroyed, from the great Sir Ernest White. If one could still track it back from here, one would find that Sir Ernest heard of it over coffee one morning from one of his research students, who had seen a proof of the result, in Swedish, in the first issue of a mathematical magazine which never produced a second issue and is not available in the libraries. And so on. Not very surprisingly, it is common for the contents of a Folk Theorem to change dramatically as its history is investigated.

I have made no mention of Special Methods such as division by zero, taking wrong square roots, manipulating divergent series, and so forth. These methods, while very powerful, are adequately described in the standard literature. Nor have I discussed the little-known Fundamental Theorem of All Mathematics, which states that every number is zero (and whose proof will give the interested reader many hours of enjoyment, and excellent practice in the use of the methods outlined above). However, it will have become apparent what riches there are in the study of Generalized Logic, and I appeal to Mathematics Departments to institute formal courses in this discipline. This should be done preferably at undergraduate level, so that those who go teaching with only a Bachelor's degree should be familiar with the subject. It is certain that in the future nobody will be able to claim a mathematical education without a firm grounding in at least the practical applications of Generalized Logic.

1 P A M Dirac, *'The Evolution of the Physicist's Picture of Nature'*, *Scientific American*, May 1963, p 47

## Basic science

ANON

From *Journal of Irreproducible Results* **13**, 5 (1964).

*Basic science*
*has to do with isotopes and ions*
*sols and gels*
*inorganic and organic smells*
*and variously differentiated cells.*
*In this scientific mélange*
*Plus c'est la même chose, plus ça change.*
*What people write*
*Was out of date on the previous night.*
*No sooner do you see data neatly analysed*
*Than BOOM comes another research*
*and the facts are changed.*

*To call this 'basic' is exaggeration.*
*Science is too ephemeral, too full of imitation.*
*A foundation or basis*
*should have homeostasis.*
*That which is basic is Art*
*of which Science is a metaplastic part.*

# On the nature of mathematical proofs

JOEL E COHEN

Condensed from
*Opus*, May 1961. Bertrand Russell has defined mathematics as the science in which
we never know what we are talking about or whether what we
are saying is true. Mathematics has been shown to apply widely
in many other scientific fields. Hence most other scientists do not
know what they are talking about or whether what they are
saying is true. Thus providing a rigorous basis for philosophical
insights is one of the main functions of mathematical proofs.

To illustrate the various methods of proof we give an example
of a logical system.

## THE PERJORATIVE CALCULUS

**Lemma 1.** *All horses are the same colour* (by induction).

**Proof.** It is obvious that one horse is the same colour. Let us
assume the proposition $P(k)$ that $k$ horses are the same colour
and use this to imply that $k + 1$ horses are the same colour.
Given the set of $k + 1$ horses, we remove one horse; then the
remaining $k$ horses are the same colour, by hypothesis. We
remove another horse and replace the first; the $k$ horses, by
hypothesis, are again the same colour. We repeat this until by
exhaustion the $k + 1$ sets of $k$ horses have each been shown to
be the same colour. It follows then that since every horse is the
same colour as every other horse, $P(k)$ entails $P(k + 1)$. But
since we have shown $P(1)$ to be true, $P$ is true for all succeeding
values of $k$, that is, all horses are the same colour.

**Theorem 1.** *Every horse has an infinite number of legs.* (Proof by
intimidation.)

**Proof.** Horses have an even number of legs. Behind they have
two legs and in front they have fore legs. This makes six legs,
which is certainly an odd number of legs for a horse. But the
only number that is both odd and even is infinity. Therefore
horses have an infinite number of legs. Now to show that this is
general, suppose that somewhere there is a horse with a finite

34

number of legs. But that is a horse of another colour, and by the lemma that does not exist.

**Corollary 1.** *Everything is the same colour.*

**Proof.** The proof of lemma 1 does not depend at all on the nature of the object under consideration. The predicate of the antecedent of the universally-quantified conditional 'For all $x$, if $x$ is a horse, then $x$ is the same colour,' namely 'is a horse' may be generalized to 'is anything' without affecting the validity of the proof; hence, 'for all $x$, if $x$ is anything, $x$ is the same colour.'

**Corollary 2.** *Everything is white.*

**Proof.** If a sentential formula in $x$ is logically true, then any particular substitution instance of it is a true sentence. In particular then: 'for all $x$, if $x$ is an elephant, then $x$ is the same colour' is true. Now it is manifestly axiomatic that white elephants exist (for proof by blatant assertion consult Mark Twain 'The Stolen White Elephant'). Therefore all elephants are white. By corollary 1 everything is white.

**Theorem 2.** *Alexander the Great did not exist and he had an infinite number of limbs.*

**Proof.** We prove this theorem in two parts. First we note the obvious fact that historians always tell the truth (for historians always take a stand, and therefore they cannot lie). Hence we have the historically true sentence, 'If Alexander the Great existed, then he rode a black horse Bucephalus.' But we know by corollary 2 everything is white; hence Alexander could not have ridden a black horse. Since the consequent of the conditional is false, in order for the whole statement to be true the antecedent must be false. Hence Alexander the Great did not exist.

We have also the historically true statement that Alexander was warned by an oracle that he would meet death if he crossed a certain river. He had two legs; and 'fore-warned is four-armed.' This gives him six limbs, an even number, which is certainly an odd number of limbs for a man. Now the only number which is even and odd is infinity; hence Alexander had an infinite number of limbs. We have thus proved that Alexander the Great did not exist and that he had an infinite number of limbs.

It is not to be thought that there are not other types of proofs, which in print shops are recorded on proof sheets. There is the bullet proof and the proof of the pudding. Finally there is 200 proof, a most potent spirit among mathematicians and people alike.

## Arrogance in physics

LAURA FERMI

Leo Szilard stayed several years with the phage group between two periods of intense political activity. Before revealing his interest in the phage, Szilard had visited Luria's laboratory at the University of Indiana. 'Doctor Szilard, I don't know how much to explain,' said Luria, embarrassed by the presence of the great nuclear physicist. 'I don't know what to assume . . .' 'You may assume,' Szilard replied promptly, 'infinite ignorance and unlimited intelligence.'

As you may know, the Institutes for Basic Research at the University of Chicago were created to continue the wartime collaboration between disciplines that had been traditionally departmentalized. Enrico was a member of the Institute for Nuclear Studies, and there was also an Institute for radiobiology. The collaboration with the biologists, Enrico once said, did not work. The trouble was that the biologists wouldn't listen to 'us'. 'Us' was, of course the physicists, and Enrico went on to explain to me that biology was in bad shape. Biologists were collecting a large number of facts, but went at it unsystematically, without a scheme or any structure to their research. Physicists could teach biologists to use the methods of physics; biologists could use the physicists' experience, and then there would be real achievements in biology. They wouldn't listen, so, too bad for them. (There was an implication of magnanimity on the part of the physicists in their willingness to pass on their experience free of any charges.)
We never took up the subject again, but the essence of this conversation stayed with me. I was puzzled by Enrico's conviction that physicists knew so much better. . . .

# What do physicists do?

From *Physicists continue to laugh* MIR Publishing House, Moscow 1968. Translated from the Russian by Mrs Lorraine T Kapitanoff.

In keeping with the spirit of the times the Editors of the wall newspaper 'Impulse' of the Physical Institute of the Academy of Science of the USSR have formed a Department of Sociological Investigations. Members of this department conducted a survey of the Moscow populace on the theme 'What do Physicists do?'

| Population Group | Total Questioned | Answered | Don't Know | Answers |
|---|---|---|---|---|
| Writer-Realists | 11 | 7 | 4 | They argue until hoarse in smoke filled rooms. It is not known why they set up unintelligible dangerous experiments using huge apparatus. |
| Writer-Visionaries | 58 | 58 | 0 | They work on enormous electronic machines called electronic brains. They work primarily in the cosmos. |
| First year college students | 65 | 65 | 0 | They speculate a lot. They make discoveries no less than once a month. |
| Graduate students | 30 | 10 | 20 | They solder circuits. They ask the older ones to find the leak. They write articles. |
| Young scientific Staff members—experimenters | 19 | 19 | 0 | They run to the equipment department. They scrub rotary vacuum pumps. They flap their ears at seminars. |
| Young scientific Staff members—theoreticians | 19 | 19 | 0 | They converse in corridors helping to make great discoveries. They write formulae, mostly incorrect. |
| Older scientific Staff members | 7 | 6 | 1 | They attend meetings. They help younger scientific staff members to find the leak. |
| Members of the personnel department | 5 | 5 | 0 | Experimenters must arrive at 8.25 so that at 8.30 they can sit silently next to apparatus which is running. Theoreticians do not work at all. |
| Members of the guard force | 6 | 6 | 0 | They walk back and forth. They present passes upside down. |
| Representatives of the Ministry of Finance | 18 | 18 | 0 | They spend money to no purpose |

# Physics terms made easy

Mostly anonymous; most of the optical terms taken from a *'Glossary of Optical Terminology'* in a volume presented to Dr Rudolf Kingslake on his retirement from the Eastman Kodak Company.

| | |
|---|---|
| Calculus of residues | How to clean up a bathtub ring |
| Catoptric | A feline eye |
| Coma | Italian: multi-toothed device for re-arranging one's hair |
| Commutator | A student who drives to school |
| Conic section | Funny paper |
| Corona | An officer who enquires into the manner of violent death |
| Cosine | The opposite of 'Stop sign' |
| Cusp | To use profane language |
| Exit pupil | A retiring student |
| Flux | Past participle of the verb 'to flex' |
| Gram | To review for examinations |
| Grand canonical ensemble | Ecumenical council |
| Graph | Principal item of a cow's diet |
| Ground state | Coffee, before brewing |
| Harmonic function | Concert |
| Hermitian operator | Recluse surgeon |
| Humbug | Noisy wiretap |
| Hypotenuse | Animal like rhinoceros but with no horn on nose |
| Len | Singular of lens, specifically a one-surface optical element |
| Marginal ray | A ray of doubtful origin |
| Millimetre | A bug like a centimetre but with more legs |
| Normal solution | The wrong answer |
| Orifice | Headquarters or place of business |
| Paradox | Two Ph D's |
| Polygon | A dead parrot |
| Poynting vector | A redundant term; all vectors point |
| Spectra | A female ghost |
| Sphere | A long, pointed weapon |
| Spin operator | Owner of a Ferris wheel |
| Statistical correlation | 36–22–35 |
| Torque | Conversation |
| Ultraviolet catastrophe | Bad sunburn |
| Vortex | Point of a mathematical figure opposite the base |
| Watt | Will you please repeat that remark? |

# Humphry Davy's first experiments

From *Collected Works of Humphry Davy* ed. John Davy (1839).

[*Among the key experiments which helped to establish that heat was a mode of motion were Humphry Davy's ice-rubbing experiments. He was aged 20 when the paper was published in 1799, working as superintendent of a Pneumatic Institution in Bristol, whose function was to establish the beneficial effects of inhaling gases. (The account of Experiment III is defective and has been slightly edited).*]

EXPERIMENT II

I procured two parallelopipedons of ice, of the temperature of $29°$, six inches long, two wide, and two-thirds of an inch thick: they were fastened by wires to two bars of iron. By a peculiar mechanism, their surfaces were placed in contact, and kept in a continued and violent friction for some minutes. They were almost entirely converted into water, which water was collected, and its temperature ascertained to be $35°$, after remaining in an atmosphere of a lower temperature for some minutes. The fusion took place only at the plane of contact of the two pieces of ice, and no bodies were in friction but ice. From this experiment it is evident that ice by friction is converted into water, and according to the supposition its capacity is diminished; but it is a well-known fact, that the capacity of water for heat is much greater than that of ice; and ice must have an absolute quantity of heat added to it, before it can be converted into water. Friction consequently does not diminish the capacities of bodies for heat.

From this experiment it is likewise evident, that the increase of temperature consequent on friction cannot arise from the decomposition of the oxygen gas in contact, for ice has no attraction for oxygen. . . .

EXPERIMENT III

I procured a piece of clock-work so constructed as to be set to work in the exhausted receiver; one of the external wheels of this machine came in contact with a thin metallic plate. (The metal of the machine and plate weighed near half a pound; on the plate was placed eighteen grains of wax.) A considerable degree of sensible heat was produced by friction between the wheel and plate when the machine worked uninsulated from bodies capable of communicating heat. I next procured a small piece of ice; round the superior edge of this a small canal was made and filled with water. The machine was placed on the ice, but not in contact with the water. Thus disposed, the whole was placed under the receiver, (which had been previously filled with carbonic

39

acid), a quantity of potash (ie caustic vegetable alkali) being at the same time introduced.

The receiver was now exhausted. From the exhaustion, and from the attraction of the carbonic acid gas by the potash, a vacuum nearly perfect was, I believe, made.

The machine was now set to work. The wax rapidly melting, proved the increase of temperature.

Caloric then was collected by friction; which caloric, on the supposition, was communicated by the bodies in contact with the machine. In this experiment, ice was the only body in contact with the machine. Had this ice given out caloric, the water on the top of it must have been frozen. The water on the top of it was not frozen, consequently the ice did not give out caloric. The caloric could not come from the bodies in contact with the ice; for it must have passed through the ice to penetrate the machine, and an addition of caloric to the ice would have converted it into water. . . . It has then been experimentally demonstrated that caloric, or the matter of heat, does not exist.

Heat, then, or that power which prevents the actual contact of the corpuscles of bodies, and which is the cause of our peculiar sensations of heat and cold, may be defined as a peculiar motion, probably a vibration, of the corpuscles of bodies, tending to separate them. It may with propriety be called the repulsive motion.

[*In 1935, Andrade turned a critical eye on these experiments which, as he remarked, have never been repeated. He was suspicious of them because 'if the ice is covered with a film of water, the friction is so small that scarcely any work is done, while if it really is dry it is liable to stick. . . . Again, the amount of work required to melt 1 gm of ice is very large . . .'.*]

From *Nature* **135** 359 (1935)

In the first of these experiments described in less than three hundred words, without any detail, Davy says that he fastened two pieces of ice by wires to two iron bars and that 'by a peculiar mechanism' the ice was kept in violent friction for some minutes. The pieces of ice 'were almost entirely converted into water' which, strangely enough, was found to be at 35° 'after remaining in an atmosphere at a lower temperature for some minutes', or, in other words, the friction of ice can raise water many degrees above the melting point! Even supposing that the stroke of the 'engine' was 5cm, and that it executed 100 strokes a minute, and that the coefficient of friction was 0·5, this would mean if for 'some minutes' we read 'ten minutes', that the force pressing the pieces of ice together would have to be equivalent to an additional pressure of about 4 atmospheres. The whole experiment is fantastic. This is said in no disrespect to Davy: how could one expect an untrained boy in 1799 to carry out an experiment which even today would tax an experienced physicist to say the least? No doubt the whole effect observed by Davy was due to conduction.

The other experiment, the one in a vacuum, was not concerned with ice at all, but with the melting of wax. The wax was apparently attached to a metal plate, against which rubbed a clockwork-driven wheel. The clockwork stood on a piece of ice in which was cut a channel containing water, and the whole was under an exhausted bell-jar. The argument was that if the heat required to melt the wax had passed from the ice to the clockwork, the water would have frozen. As, however, the heat required to produce the rise of temperature observed in the clockwork amounted to but 12 calories, only 0·15 cc of water would have frozen in any event, which actually could not be observed by eye in a rough channel cut in a piece of ice. The experiment proves nothing at all.

I may be held to have spent too much time on a point which some may say is of historical interest only. I hold, however, that it is very inadvisable that students should be taught to attach a fundamental importance, not to experiments crudely carried out, which were afterwards improved, but to experiments of which one probably cannot be carried out at all, while the other is so ill-designed as to prove nothing. I am no denigrator; I do not think that it detracts from the greatness of Davy to point out that his first experiments, carried out when he was a country lad, were uncritical and lacked all quantitative basis. It is time, however, that they ceased to be ranked with such convincing demonstrations as those of Rumford, and disappeared from the textbooks. Or, if they are quoted, do let us have instructions as to how to melt two pieces of ice by rubbing them together in a vacuum.

[*Davy's paper had a real and lasting effect on the development of science. But do not, Gentle Young Reader, publish as hastily as Davy unless you are sure that you are as great a genius as he.*]

Basic research is what I am doing when I don't know what I am doing. *Werner von Braun*

# Maxwell's aether

From *Philosophical Magazine* [4] **21**, 281 (1861).

*[Nowadays, we present electromagnetic theory in an abstract way, but this was not the method of the innovators. Maxwell began by making a model of the aether, composed of vortices of sub-molecular size, all rotating in the same direction so as to produce the circulation of the magnetic field. In the spirit of the age he took the model very seriously.]*

I have found great difficulty in conceiving of the existence of vortices in a medium, side by side, revolving in the same direction about parallel axes. The contiguous portions of consecutive vortices must be moving in opposite directions; and it is difficult to understand how the motion of one part of the medium can coexist with, and even produce, an opposite motion of a part in contact with it.

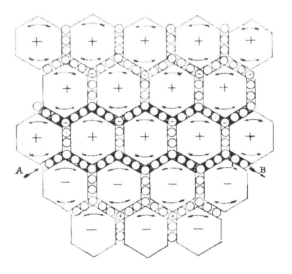

The only conception which has at all aided me in conceiving of this kind of motion is that of the vortices being separated by a layer of particles, revolving each on its own axis in the opposite direction to that of the vortices, so that the contiguous surfaces of the particles and of the vortices have the same motion.

In mechanism, when two wheels are intended to revolve in the same direction, a wheel is placed between them so as to be in gear with both, and this wheel is called an 'idle wheel'. The hypothesis about the vortices which I have to suggest is that a layer of particles, acting as idle wheels, is interposed between each vortex and the next, so that each vortex has a tendency to make the neighbouring vortices revolve in the same direction with itself.

*[The idle wheels were particles of electricity. He deduced the electromagnetic equations on the basis of this picture.]*

It serves to bring out the actual mechanical connexions between the known electro-magnetic phenomena; so that I venture to say that any one who understands the provisional and temporary character of this hypothesis, will find himself rather helped than hindered by it in his search after the true interpretation of the phenomena.

## Boltzmann on style in physics

From Ludwig Boltzmann's contribution to Kirchhoff's *Festschrift*, 1888.

Just as a musician recognizes Mozart, Beethoven or Schubert from the first few bars, so does a mathematician recognize his Cauchy, Gauss, Jacobi or Helmholtz from the first few pages. Perfect elegance of expression belongs to the French, the greatest dramatic vigour to the English, above all to Maxwell. Who does not know his dynamical theory of gases? First, majestically, the Distribution of Velocities develops, then from one side the Equations of Motion in a Central Field; ever higher sweeps the chaos of formulae; suddenly are heard the four words: 'Put $n = 5$'. The evil spirit $V$ (the relative velocity of two molecules) vanishes and the dominating figure in the bass is suddenly silent; that which had seemed insuperable being overcome as if by a magic stroke. There is no time to say why this or why that substitution was made; he who cannot sense this should lay the book aside, for Maxwell is no writer of programme music obliged to set the explanation over the score. Result after result is given by the pliant formulae till, as unexpected climax, comes the Heat Equilibrium of a heavy gas; the curtain then drops.

One of R W Wood's associates told me of the time he was sent a collaborator from Rutherford's laboratory. The visitor was found to be an ideal colleague and Wood wanted to keep him at Johns Hopkins, but Rutherford decreed differently. The man left for home and Wood sighed, 'The Lord giveth and the Lord taketh away.'

PAUL KIRKPATRICK

# An Experiment *to prove, that* Water, *when agitated by* Fire, *is infinitely more* elastic *than* Air *in the same* Circumstances; *by the late Rev$^d$* John Clayton, *Dean of* Kildare *in* Ireland

Condensed from *Philosophical Transactions of the Royal Society* 41, 162–66 (1739).

[*Scientific writing in the eighteenth century was uninhibited and direct, a contrast to the dullness of much that fills today's journals. This article is a typical example of early writing. The author describes how he found that the vapour pressure (the 'elasticity') of water increases very much more rapidly than the pressure of air as the temperature is raised.*]

Sir *Thomas Proby* having heard of a new Digester, which I contrived, had a Desire to see it, and some Experiments made therein. I had a small one, which I designed only for an inward Cylinder; this I could easily put in my Pocket: Wherefore, going to pay him a Visit at *Elton* in *Huntingdonshire*, I took it along with me; and having softened a Bone therein in a very short Space, he was desirous to know the shortest Time it was possible to soften a Bone in: I told him, I thought I could soften the Marrow-bone of an Ox in a very few Minutes, but that the Vessel was very weak, and I feared would not endure the Pressure of so violent a Heat; yet seeming desirous to have the Experiment tried, I said I was ready to venture my Vessel: Then having fixed all things right, and included about a Pint of Water, and I believe, about 2 oz of a Marrow-bone, we placed the Vessel horizontally betwixt the Bars of the iron Grate into the Fire about half way; and in three Minutes time I found it raised to a great Heat; whereupon I had a mind to have it taken out of the Fire, lest it should have burst; telling Sir *Thomas* of the Danger that I apprehended. Scarce had I done speaking, and Sir *Thomas* thereupon moved his Chair to avoid Danger; but seeing the Heat becoming more raging, I stepped to the Side-table for the Iron wherewith I managed the Digester, in order to take it out of the Fire, when, on a sudden, it burst as if a Musquet had gone off. A Maid that was gone a milking, heard it at a considerable Distance; the Servants said it shook the House. The Bottom of the Vessel, that was in the Fire, gave way; the Blast of the expanded Water blew all the Coals out of the Fire all over the Room. All the Vessel together flew in a direct Line cross the Room, and hitting the Leaf on a Table made of an inch Oak plank, broke it all in Pieces, and rebounded half way of the Room back again. What surprised me in this Event was, that the Noise it made at its bursting was by no means like the successive evaporating of an Æolipile, but like the firing off of Gunpowder. Nor could I perceive anywhere in the Room the least Sign of Water, though I looked carefully for it, and, as I said before, I had put a Pint into the Digester, save only that the Fire was quite extinguished, and every Coal belonging to it was black in an Instant.

But to confirm the Elasticity of Water, or to shew at least, that there is a much stronger elastic Force in Water and Air, when jointly included in a Vessel, than when Air alone is inclosed therein, I made the following Experiment: I took two 6 oz Phials, into the one I put about 5 oz of Water, or better, and so corked it as well as I possibly could; the other I corked in the same Manner, without putting any thing into it. I inclosed them both in my new Digester, Four-fifths being filled with Water; when the Heat was raised to about Five-seconds, I heard a considerable Explosion, and a jingling of Glass within the Vessel, and shortly after another Explosion, but not so loud as the former; whence I concluded, that both the Phials were broken. I then let the Digester cool leisurely, and the next Day I opened it; both the Corks were swimming on the Top of the Water, but only one of the Phials was broken, *viz*. that one into which I had not put any Water. At first, indeed, I concluded, that the Pressure or Dilatation of the Air in the empty Phial being stronger than the ambient Pressure, forced forth the Cork, whereupon the Water, rushing in the Violence, might break the Phial; and therefore that this was the Cause also of the Loudness of the Explosion; whereas the other being mostly filled with Water, there being but a small Quantity of Air therein, just enough to force out the Cork, the Phial was not broken, but was preserved by the Force of the Water inclosed therein. But I have had Reason since to change my Opinion; for having had very strong Phials made on Purpose to make some peculiar Experiments therewith, I took one of them, and having filled it about a quarter full with Water, and corked it very well, I set it in a square iron Frame, with a Screw to screw down the Cork, and keep it from flying forth. I then put it into a Digester, Four-fifths filled with Water; which being heated to a due Height, when I opened it, I found the Cork forced into the Phial, though the Cork was so very large, that it amazed several who saw it, to conceive how it was possible for so large a Cork to be forced into the Bottle. Hence it manifestly appears, that the Pressure in the Digester, wherein was proportionately more Water, and less Air, was stronger than the Pressure within the Phial, wherein was proportionately more Air, and less Water. Then I reasoned thus also of the two former Phials: That the Air in the Phial, wherein was no Water included, making not

*An* Experiment *to prove, that* Water, *when agitated by* Fire, *is infinitely more* elastic *than* Air *in the same Circumstances*

a proportionate Resistance to the ambient Pressure in the Digester, wherein was a considerable Quantity of Water, the Cork was forced inward with such Violence, that it, together with the Water, dashed the Phial in pieces; but that in the other Phial, wherein there were Five-sixths of Water, the inward Pressure in the Phial being greater than the ambient Pressure in the Digester, wherein were but Four-fifths of Water, the Cork was thereby forced outward; and that the small Difference between the proportionate Quantity of Water and Air in the Phial and in the Digester, being only a Four-fifth to Five-sixths, was the Reason not only why the Bottle was not broken, but also of the Faintness of the Explosion.

From *The Space Child's Mother Goose*, verse by Frederick Winsor, illustration by Marion Parry (New York: Simon and Schuster) 1958.

*Three jolly sailors from Blaydon-on-Tyne*
*They went to sea in a bottle by Klein.*
*Since the sea was entirely inside the hull*
*The scenery seen was exceedingly dull.*

## H A Rowland

PAUL KIRKPATRICK

I have been told of an incident in the life of H A Rowland which I cannot certify but which is consistent with other reports of his personality. It seems he was called on to testify as a science expert in some kind of court case. In exploring his competence an attorney asked him who was the foremost American physicist. Unhesitatingly, Rowland answered, 'I am.' Later a friend reproached him gently for his immodesty. Rowland's response was, 'Well, you have to remember I was under oath.'

Dial Barometers, 8–14 inch, plain carved in solid oak, mahogany, rosewood or walnut frames with double basil ring and polished-edge plate glass, or richly carved in oak, mahogany or walnut wood with solid frames of Gothic, Medieval, Elizabethan, Egyptian, Chippendale or other designs. Price £5-5s to £25. Suitable for Club Houses, Mansions, etc.

(From *Negretti and Zambra's Encyclopaedic Illustrated and Descriptive Reference Catalogue of Optical, Mathematical, Physical, Photographic and Standard Meteorological Instruments, manufactured and sold by them.*)

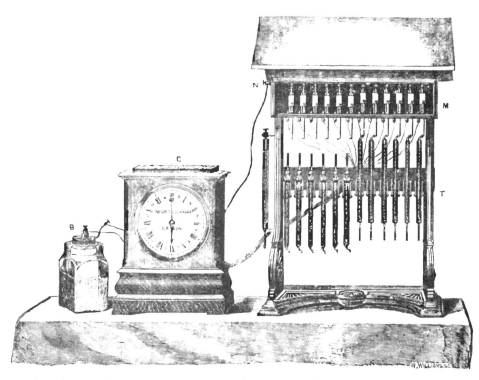

A Victorian solution to the meteorologist's problem of reading the air temperature at hourly intervals. Each mercury-in-glass thermometer T in its housing is turned bulb downwards and held in that position by a spring-loaded stop or *detent*. At a certain moment the spring is released by its electromagnet M, the thermometer falls upside down, the mercury thread breaks and the temperature reading is preserved. Of the twelve thermometers, the six on the left have fallen, the six on the right have not yet been actuated. Presumably the attendant will come in another six hours to read the whole row and record the results.

(From *Negretti and Zambra's Encyclopaedic Illustrated and Descriptive Reference Catalogue of Optical, Mathematical, Physical, Photographic and Standard Meteorological Instruments, manufactured and sold by them*.)

# Confrontation

MAURICE CAULLERY and ANDRÉE TÉTRY

From *A General History of the Sciences: Science in the Nineteenth Century* ed René Taton, translation by A J Pomerans (London: Thames and Hudson 1966) pp 477–8.

It is not easy today to realize the fierceness with which this war between science and religion, centred about evolution, was carried on in the late 19th century, but the bitterness of the controversy may be illustrated by the attempt made by Sir Richard Owen and Bishop Wilberforce to reduce Darwin's theory to ridicule at the British Association meeting at Oxford, in June 1860. At the Thursday (28th June) meeting Sir Richard Owen made the rash statement that a gorilla's 'brain is more different from a man's brain than it is from the brain of the lowest apes,' and Huxley at once 'flatly and unequivocally' contradicted the statement. By the Saturday, when Wilberforce was booked to speak, feelings were running high. Orthodox churchmen everywhere felt that the Christian religion, whose doctrines were based on a literal interpretation of the Bible, was being threatened by the arrogance of science, and Wilberforce openly told his friends that he now intended 'to smash Darwin'. The zealous bishop had not himself troubled to read Darwin's *Origin of Species*, but he had been coached in his part by Sir Richard Owen.

He spoke eloquently to a packed meeting, there being so many people present that even the window-sills were occupied. Carried away by his enthusiasm, he turned at one point to Huxley and, with a fine show of scorn, asked him, 'Is it on your grandfather's or your grandmother's side that the ape ancestry comes in?' His final conclusion that 'Darwin's theory is contrary to the revelations of God in the Scriptures' was greeted with wild cheering, and there the meeting would have ended had not many of the students clamoured for a reply from Huxley. At length Huxley rose and made his celebrated retort: 'I asserted, and I repeat, that a man has no reason to be ashamed of having an ape for his grandfather. If there were an ancestor whom I should feel shame in recalling it would be a man of restless and versatile intellect who, not content with success in his own sphere of activity, plunges into scientific questions with which he has no real acquaintance, only to obscure them by aimless rhetoric and distract the attention of his hearers from the point at issue by digressions and appeals to religious prejudice.'

Thus, the bishop was put in his place, and though many people present were shocked and Lady Brewster, among other notable women, thought it proper to faint, the final consensus of opinion was overwhelmingly on the side of science and the great conception of evolution.

# Getting bubble chambers accepted by the world of professional physicists

DONALD A GLASER

[*Awarded the 1960 Nobel Prize for Physics for his invention of the bubble chamber.*]

My first talk on the subject was scheduled by the secretary of the American Physical Society, Karl Darrow, in the Saturday afternoon 'crackpot session' of the American Physical Society. My first paper on this subject was returned as unaccepted (*Phys. Rev. Lett.*) on the grounds that I had used the word 'bubblet' which is not in Webster. I was refused support by ONR, AEC, and NSF on the grounds mentioned by one agency that 'the work was too speculative to spend government funds for it,' and I was denied access to the Cosmotron on the same grounds. These were not painful experiences because the University of Michigan found $750 to support my research for a year which was all that I really needed during that first year.

One night during a summer physics colloquium extending over a week or two attended by Bruno Rossi, Chandresekhar, Uhlenbeck and a number of others, we were sitting around drinking beer in the local student pub when one of the men staring dreamily into the pitcher of beer on the centre of the table said, 'Glaser, bubble chambers ought to be really easy, you can see tracks in damn near anything.' All the teasing was good-natured, however, and the story had a happy ending.

---

## Bunsen burner

HENRY ROSCOE

From *Bunsen Memorial Lecture, Journal of the Chemical Society* 77 (1900).

When he first came to Heidelberg in the summer of 1852, Bunsen found himself installed in Gmelin's old laboratory. This was situated in the buildings of an ancient monastery, and there we all worked. It was roomy enough; the old refectory was the main laboratory, the chapel was divided into two, one half became the lecture-room and the other a storehouse and museum. Soon the number of students increased and further extensions were needed, so the cloisters were enclosed by windows and working benches placed below them. Beneath the stone floor at our feet slept the dead monks, and on their tomb-stones we threw our waste precipitates! There was no gas in Heidelberg in those days; nor any town's water supply. We worked with Berzelius's spirit lamps, made our combustions with charcoal, boiled down our washwaters from our silicate analyses in large glass globes over charcoal fires, and went for water to the pump in the yard.

Some short time before the opening of the new laboratory, in 1855, the town of Heidelberg was for the first time lighted with gas, and Bunsen had to consider what kind of gas-burner he would use for laboratory purposes. Returning from my Easter vacation in London, I brought back with me an Argand burner with copper chimney and wire-gauze top, which was the form commonly used in English laboratories at that time for working with a smokeless flame. This arrangement did not please Bunsen in the very least; the flame was flickering, it was too large, and the gas was so much diluted with air that the flame-temperature was greatly depressed. He would make a burner in which the mixture of gas and air would burn at the top of the tube without any gauze whatsoever, giving a steady, small, and hot, non-luminous flame under conditions such that it not only would burn without striking down when the gas supply was turned on full, but also when the supply was diminished until only a minute flame was left. This was a difficult, some thought it an impossible, problem to solve, but after many fruitless attempts, and many tedious trials, he succeeded, and the 'Bunsen burner' came to light.

Illustration by courtesy of Negretti and Zambra Ltd.

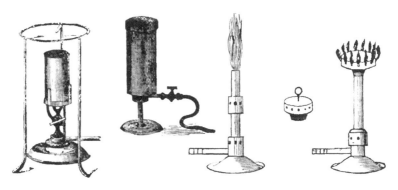

Report writing, like motor-car driving and love-making, is one of those activities which almost every Englishman thinks he can do well without instruction. The results are of course usually abominable.

TOM MARGERISON, reviewing *'Writing Technical Reports'*— by Bruce M Cooper, in the *Sunday Times* 3 January 1965

# Rutherford and Nature's whispers

A S RUSSELL

From '*Lord Rutherford: Manchester*', 1907–19: a partial portrait by A S Russell in *Rutherford at Manchester* ed J B Birks (London: Heywood) 1962.

Rutherford's great gift was to design experiments that asked of Nature the most pertinent questions and then to brood for long over the answers. In this respect he was of the great company of Newton and Faraday. They knew what to ask and how to pay attention not so much to what Nature was saying as to what Nature was whispering. In this Rutherford was an artist. All his experiments had style. Let me illustrate: One of the early experiments he did at Manchester was with Royds on the identification of the $\alpha$-particle with the atom of helium. He had known for years that the $\alpha$-particle was likely to be the helium atom, but he had to make assurance doubly sure. A glass tube blown so thin that it allowed the $\alpha$-particle easily to penetrate its walls was shown to be gas-tight. Filled with radon it was surrounded by a second glass tube highly evacuated. In this tube it was simple to show that helium accumulated with the passage of time as it got filled by the particles entering it. Then, in 1908, how beautiful, as well as how accurate, was Rutherford's determination of Avogadro's number! He counted accurately the number of $\alpha$-particles emitted in a given time by a known mass of radium and determined also the value of the charges the particles were carrying. From these data he obtained a value of Avogadro's number which was 40 per cent different from the best of earlier determinations but which is still within 3 per cent of the best determinations of today. Or again, think of the simplicity of the device developed by Geiger and himself to register a single $a$-particle. A wire charged almost to breaking potential and connected to an electrometer was inserted in a tube into which a very small stream of $a$-particles was allowed to enter. As one particle entered, its feeble ionization increased greatly by the ionization from collision, sufficed to cause a discharge easily registered by an electrometer. Or think of the work on the scattering of $\alpha$-particles by thin films of metal where the number scattered through a given angle was counted by well-rested eyes in a dark room by the flashes which each one individually gave on a screen of zinc sulphide; or the comparison of two very disparate standards of $\gamma$-radiation by putting each in turn on an optical bench at such distances from a measuring instrument that a constant result was recorded, and then invoking the inverse square law for the calculation. On a backward view one saw the beauty of the method of investigation as well as the ease with which the truth was arrived at. The minimum of fuss went with the minimum of error. With one movement from afar Rutherford, so to speak, threaded the needle first time.

# The organization of research—1920

WILLIAM MORTON WHEELER

Condensed from the address of the retiring Vice President and Chairman of the section for Zoological Sciences, American Association for the Advancement of Science, Chicago 1920. Reprinted in *Science* 53, 53 (1920).

For this address I selected the most fashionable and exalted topic I could find, for you must all have observed that at the present time no word occurs with greater frequency and resonance in serious discourse than 'organization.' Everybody is so busy organizing something that the word's subtly concealed connotations of control and regulation appear to be overlooked. The purpose of organization is instrumental, as is shown by the derivation of the word, from 'organon' or tool, or implement, which is in turn derived from 'ergo' to work. It is one of those superb, rotund words which dazzle and hypnotize the uplifter and eventually come to express the peculiar spirit or tendency of a whole period.

These words, which for want of a better term I may call 'highbrow,' and the conceptions they embody, are so interesting that I will dwell on them for a moment. During the late Victorian period the most highbrow word was 'progress.' It disappeared and gave place to organization with the World War when we realized that the evolution of our race since the Neolithic Age was not nearly as substantial as we had imagined. Neither the Greeks nor the people of the Middle Ages seem to have had either of these words or their conceptions, though the Greeks, at least, did a fair amount of progressing and organizing. The Mediaeval highbrow words were 'chivalry' and 'honour,' the latter persisting down to the present day in Continental Europe in the German students' duelling code, a living fossil . . . Schopenhauer remarked that the duel and venereal diseases were the only contributions to culture the race had made since the classical period, overlooking the fact that the Greeks and the Japanese had their own highbrow words and institutions. Gilbert Murray has shown that the word 'aidos,' which the Achaean chiefs of the Homeric age so solemnly uttered, was applied to a peculiar kind of chivalry, and the 'bushido' of the Japanese was another similar though independent invention. All of these conceptions—progress, organization, chivalry, aidos, bushido—seem to start among the intellectual aristocracy and all imply a certain 'noblesse-oblige,' for there is no fun in continually exhorting others to progress unless you can keep up with the procession, or of organizing others unless you yearn to be organized yourself, just as there is no fun in getting up a duelling or bushido code unless you are willing to fight duels or commit harakiri whenever it is required by the rules of the game.

Of course, the vogue of 'organization' was abnormally stimulated by the mobilization of armies and resources for the World War. We acquired the organizing habit with a vengeance and have not since had time to reflect that there may be things in the world that it would be a profanation to organize—courtship, for example—or things not worth organizing—a vacuum, for example —or things than can not be organized, or if organizable, better left as they are—scientific research, perhaps.

The outline of this paper came to me, probably after prolonged subconscious incubation, while I was wondering how much coal I would save by using as an *ersatz* the literature received during the past three years from that noble superorganization of super-organizers, the National Research Council. . . .

## Solar eclipse

REINHOLD GERHARZ

Self-appointed protectors of the people managed to persuade thousands of would-be observers in the totality path of the March 7, 1970 solar eclipse into relinquishing the only chance of their life to experience the beauty and awe of this rare event.

Our group set up instruments in the peanut farming belt of North Carolina. During the preparations, we met many local farmers who had been so frightened by the eclipse warnings of the news services that they swore not only to keep their TV-sets turned off but also to hide themselves inside their houses on account of the 'dangerous radiation' from the eclipsed sun. Not even the assurance and testimony of our own personal vigour and intents would change their minds, although I did note some bewilderment and faint relaxation after they realized our deter-mination to stay outside, and our audacity in daring all the deadly dangers which had been prophesied while the heavenly crisis lasted.

In retrospect, and after realizing that about $\$10^{12}$ have been spent on the restitution of education and science in the wake of the Sputnik Effect, I find myself amazed that 12 years of the Big Science Craze were still not sufficient to obliterate the credibility gaps between the scientists, the news media, and the rest of the populace of this nation.

# How Newton discovered the law of gravitation

JAMES E MILLER

Condensed from
'How Newton
Discovered the Law
of Gravitation' by
James E. Miller,
The American
Scientist, 39 no 1,
1951.

Few are familiar with the details of Newton's twenty-year search for a proof of his hypothesis: the frustrations and failures, the need for accurate measurements of the earth's radius and for a mathematical tool that Newton himself was forced to invent, and the integration of his scattered efforts by the splendid organization of the Fruit-Improvement Project. These details have been collected from his *Principia*, personal letters, notebooks, and other papers, and a series of personal interviews arranged by a medium of the author's acquaintance.

In 1665 the young Newton became a professor of mathematics in the University of Cambridge, his *alma mater*. His services to his college went far beyond the mere act of classroom teaching. He was an able and active member of the college's curriculum committee, the board of the college branch of the Young Noblemen's Christian Association, the Dean's advisory committee on scholarships and awards, the committee for discipline, the ground committee, the publications committee, the *ad hoc* committee, and numerous other committees essential to the proper administration of a college in the seventeenth century. An exhaustive compilation of Newton's work along these lines reveals that, during a five-year period, he served on 379 committees, which investigated an aggregate of 7924 problems of campus life and solved 31 of them.

His unselfish devotion to the important work of his committees absorbed so much time that he was constrained to turn more and more of his teaching duties over to one of his students. He reasoned, quite correctly, that the substitution of a student as teacher in his place would benefit both the student and the student's students: the former because, in teaching, his own knowledge would be enhanced; and the latter because, in being taught by one near to them in age and interests, they would more eagerly grasp at the scraps of knowledge that came their way.

At about this time Newton, whose mind was too active ever to let scientific problems recede from his attention, occasionally mulled over the great discoveries of Kepler on planetary motions and the hypothesis advanced by a number of astronomers, that these motions were governed by an attraction that varied inversely as the square of the distance between planets. One evening of a crowded day in the year 1680, a committee that was scheduled to meet at eleven o'clock, no earlier time being available, was unable to muster a quorum because of the sudden death from exhaustion of one of the older committee members. Every waking moment of Newton's time was so carefully budgeted that he found himself with nothing to do until the next committee meeting at midnight. So he took a walk—a brief stroll that altered the history of the world.

It was on this excursion into the night air of Cambridge that Newton chanced to see a particularly succulent apple fall to the ground. His immediate reaction was typical of the human side of this great genius. He climbed over the garden wall, slipped the apple into his pocket, and climbed out again. As soon as he had passed well beyond that particular garden, he removed the apple from his pocket and began munching it. Then came inspiration. Without prelude of conscious thought or logical process of reasoning, there was suddenly formed in his brain the idea that the falling of an apple and the motion of planets in their orbits may be governed by the same universal law. Before he had finished eating the apple and discarded the core, Newton had formulated his hypothesis of the universal law of gravitation.

In the following weeks Newton's thoughts turned again and again to his hypothesis. Rare moments snatched between the adjournment of one committee and the call to order of another were filled with the formulation of plans for testing the hypothesis. Eventually, after several years during which, according to evidence revealed by diligent research, he was able to spend 63 minutes and 28 seconds on his plans, Newton realized that the proof of his hypothesis would take more spare time than might become available during the rest of his life. He had to find accurate measurements of a degree of latitude on the earth's surface, and he had to invent the calculus.

Finally he concluded that he must find some relief from his collegiate administrative burdens. He knew that it was possible to get the King's support for a worthy research project of definite aims, provided a guarantee could be made that the project would be concluded in a definite time at a cost exactly equal to the amount stipulated when the project was undertaken. Lacking experience in such matters, he adopted a commendably simple approach and wrote a short letter of 22 words to King Charles, outlining his hypothesis and pointing out its far-reaching implications if it should prove to be correct.

Eventually, Newton's letter and the bulky file of comments it had gathered on its travels reached the office of the secretary of HMPBRD/CINI/SSNBI—His Majesty's Planning Board for Research and Development/Committee for Investigation of New

Ideas/Subcommittee for Suppression of Non-British Ideas. The secretary immediately recognized its importance and brought it before the subcommittee, which voted to ask Newton to testify.

Newton's testimony before HMPBRD/CINI is recommended to all young scientists who may wonder how they will comport themselves when their time comes. His college considerately granted him two months' leave without pay while he was before the committee, and the Dean of Research sent him off with a joking admonition not to come back without a fat contract. The committee hearing was open to the public and was well attended, though it has been suggested that many of the audience had mistaken the hearing room of HMPBRD/CINI for that of HMCE-VAUC—His Majesty's Committee for the Exposure of Vice Among the Upper Classes.

After Newton was sworn to tell the truth and had denied that he was a member of His Majesty's Loyal Opposition, had ever written any lewd books, had travelled in Russia, or had seduced any milkmaids, he was asked to outline his proposal. In a beautifully simple and crystal-clear ten-minute speech, delivered extemporaneously, Newton explained Kepler's laws and his own hypothesis, suggested by the chance sight of an apple's fall. At this point one of the committee members, an imposing fellow, demanded to know if Newton had a means of improving the breed of apples grown in England. Newton began to explain that the apple was not an essential part of the hypothesis, but he was interrupted by a number of committee members, all speaking at once in favour of a project to improve apples. This discussion continued for several weeks, while Newton returned to his college and his important committee work.

Several months later Newton was surprised to receive a bulky package from HMPBRD/CINI. He opened the package and found it contained a variety of government forms, each in quintuplicate. His natural curiosity, the main attribute of the true scientist, provoked him into a careful study of the forms. After some time he concluded that he was being invited to submit a bid for a contract for a research project on the relationship between breed, quality, and rate of fall of apples. The ultimate purpose of the project, he read, was to develop an apple that not only tasted good but also fell so gently that it was not bruised by striking the ground. Now, of course, this was not what Newton had had in mind when he had written his letter to the King. But he was a practical man and he realized that, in carrying out the proposed project, he could very well test his hypothesis as a sort of sideline or by-product. Thus, he could promote the interests of the King and do his little bit for science in the bargain.

Having made his decision, Newton began filling out the forms without further hesitation. A true believer in correct administrative procedures, Newton sent the completed forms by special messenger to the Dean of Research, for transmittal through proper channels to HMPBRD/CINI. His adherence to established procedure was rewarded a few days later when the Dean of Research summoned him and outlined a new plan, broader in scope and more sweeping in its conception. The Dean pointed out that not only apples but also cherries, oranges, lemons, and limes fell to earth, and while they were about it they might as well obtain a real, man-sized government contract to cover all the varieties of fruit that grow above the ground. Newton started to explain the misunderstanding about the apples but he stopped rather than interrupt the Dean, who was outlining a series of conferences he proposed to organize among fruit growers and representatives of various departments of His Majesty's Government. The Dean's eyes began to glaze as he talked, and he became unaware that anybody else was in the room. Newton had an important committee meeting at that time, so he quietly went out the door, leaving the Dean of Research in an ecstasy of planning.

The season passed, while Newton led a busy, useful life as a member of many committees and chairman of some. One dark winter's day he was called again to the office of the Dean of Research. The Dean was beaming. The project was to be supported by no less than five different branches of His Majesty's Government plus a syndicate of seven large fruit growers. Newton's part in the project was to be small but important: he was to direct the Subproject for Apples.

The following weeks were busy ones for Newton. Though relieved from his committee work (a young instructor of Greek, Latin, history and manual training took his place on the committees), he found himself cast into a morass of administrative problems. He personally filled out 7852 forms, often in quintuplicate and sextuplicate; he interviewed 306 serving wenches and hired 110 of them as technical assistants. With his own hands he cleaned out an abandoned dungeon in a nearby castle for use as subproject headquarters; and, turning carpenter in typically versatile fashion, he erected twelve temporary buildings to house his staff. These buildings, used today as classrooms, stand as a monument to Newton's career.

This period of his life was a happy and profitable one for Newton. From the time he rose in the morning until, exhausted with honest labour, he dropped late at night back into his humble bed of straw, he spent each day filling out payroll forms for his serving wenches, ordering pens and paper, answering the questions of the financial office, and showing distinguished visitors and the Dean of Research around his subproject. Each week he wrote out a full progress report which was duplicated and sent by special messenger to 3388 other projects sponsored by His Majesty's Government throughout the British Isles.

One of these remarkable documents, in an excellent state of preservation, can be found in the Museum of the Horticultural Society of West Wales. In typically logical style, the report, bound in a dark red stiff cover bearing the project number, HM2wr 3801-g(293), stamped in gold leaf, opens with a succinct table of contents:

1. Administration
2. Conferences
3. Correspondence
4. Supplies
5. Results of research

The last section, 'Results of research,' may have been lost during the intervening years, or it may not have been specifically required under the terms of His Majesty's contract of that era. At any rate it is not there. But the other sections remain to gladden the hearts of those permitted to read them.

One day in 1685 Newton's precise schedule was interrupted. He had set aside a Tuesday afternoon to receive a committee of vice presidents of the fruit growers' syndicate when, much to his horror and Britain's deep sorrow, the news spread that the whole committee had been destroyed in a three-stagecoach smashup. As once before, Newton found himself with a hiatus. He took a leisurely walk through the luscious vineyards of the Subproject on Grapes, but, not, of course, until he had obtained security clearance at the gate. While on this walk there came to him, he knew not how ('Ye thought just burst upon me,' he later wrote) a new and revolutionary mathematical approach which, in less time than it takes to tell about it, could be used to solve the problems of attraction in the neighbourhood of a large sphere. Newton realized that the solution to this problem provided one of the most exacting tests of his hypothesis; and furthermore, he knew, without need of pen and paper to demonstrate the fact to himself, that the solution fully supported his hypothesis. We can well imagine his elation at this brilliant discovery; but we must not overlook his essential humility, which led him forthwith to kneel and offer thanks to the King for having made the discovery possible.

On his return from this walk, Newton stopped for a moment to browse in a bookstore, where he accidentally knocked a book to the floor. With apologies to the proprietor, Newton retrieved the book and dusted it off. It was Norwood's *Sea-Man's Practice*, dated 1636. Opening the book at random, Newton found it contained the exact information on the length of a latitude degree that he required for the complete test of his hypothesis. Almost instantaneously, one part of his brain performed several lightning calculations and presented the result for the other part to examine; and there it was: the proof complete and irrefutable. Newton glanced at the hourglass and with a start remembered that he was due back at the dungeon to sign the serving wenches' time slips as they checked out for the day. He hurried out of the bookshop.

Thus it was that His Majesty's Government supported and encouraged Newton during the trying years in which he was putting his hypothesis to the test. Let us not dally with the story of Newton's efforts to publish his proof, the misunderstanding with the editor of the *Horticultural Journal*, the rejections by the editors of the *Backyard Astronomer* and *Physics for the Housewife*, suffice it to say that Newton founded his own journal in order to make sure that his proof would be published without invalidating alterations. Regrettably, he named his journal *Star and Planet*, with the result that he was branded a subversive since Star could mean Red Star and Planet could mean Plan-It. Newton's subsequent testimony before the Subcommittee for Suppression of Non-British Ideas remains a convincing demonstration of the great qualities that combined to make him a genius. Eventually he was exonerated, and after enjoying many years of the fame that was due him, reigning one day each year as King of the Apple Festival, Newton died happily.

# Graduate students

P M S BLACKETT

From 'Memories
of Rutherford' in
Rutherford at
Manchester ed
J B Birks
(London: Hey-
wood) 1962.

Even in the Cavendish period when apparatus was inevitably getting more complicated, Rutherford could be disconcertingly unappreciative of experimental and constructional difficulties. I can confirm from personal experience what others have said, that Rutherford took only a minimal interest in one's work during these years of laborious constructional work: indeed, he was often so impatient for results that the young research student had often to exert some will-power to resist being unduly hurried. His own main personal results had been achieved with apparatus of exquisite simplicity—a simplicity arising both from his genius and from the nature of his chosen work—and he was slow to admit that these days were over for the time being, and complexity of apparatus was likely to grow and grow.

Once physical results arose from a student's experiment, then Rutherford became the stimulating and genial visitor to one's room. Rutherford's main role in these later Cavendish days (when, of course, he was already a man of affairs with many calls on his time) was to give the new student a fertile problem, leave him to it for a year or two, ignore all the years of travail, but welcome the eventual results with enthusiasm. It is surprising how well the method worked.

Rutherford once said he had never given a student a dud problem! Napoleon is reported to have once said, 'There are no bad soldiers, only bad generals'. Rutherford might have adapted this remark to some of his colleagues (and I think he certainly would have, if he had thought of it, for he had a sharp tongue particularly in the dark room while counting scintillations), 'There are no bad research students, only bad professors'.

## Epigrams

ALEXANDER POPE

*Nature and Nature's laws lay hid in night.*
*God said,* Let Newton be! *and all was light.*

SIR JOHN COLLINS SQUIRE

*It did not last: the Devil howling* 'Ho!
Let Einstein be! *restored the status quo.*

# Take away your billion dollars—1946

ARTHUR ROBERTS

[*Written while the Brookhaven National Laboratory was being planned*]

*Upon the lawns of Washington the physicists assemble,*
*From all the land are men at hand, their wisdom to exchange.*
*A great man stands to speak, and with applause the rafters tremble.*
*'My friends,' says he, 'You all can see that physics now must change.*
*Now in my lab we had our plans, but these we'll now expand,*
*Research right now is useless, we have come to understand.*
*We now propose constructing at an ancient Army base,*
*The best electronuclear machine in any place.—Oh*

*'It will cost a billion dollars, ten billion volts 'twill give,*
*It will take five thousand scholars seven years to make it live.*
*All the generals approve it, all the money's now in hand,*
*And to help advance our program, teaching students now we've banned.*
*We have chartered transportation, we'll provide a weekly dance,*
*Our motto's integration, there is nothing left to chance.*
*This machine is just a model for a bigger one, of course,*
*That's the future road for physics, as I hope you'll all endorse.'*

*And as the halls with cheers resound and praises fill the air,*
*One single man remains aloof and silent in his chair.*
*And when the room is quiet and the crowd has ceased to cheer,*
*He rises up and thunders forth an answer loud and clear:*
*'It seems that I'm a failure, just a piddling dilettante,*
*Within six months a mere ten thousand bucks is all I've spent,*
*With love and string and sealing wax was physics kept alive,*
*Let not the wealth of Midas hide the goal for which we strive.—Oh*

*'Take away your billion dollars, take away your tainted gold,*
*You can keep your damn ten billion volts, my soul will not be sold.*
*Take away your army generals; their kiss is death, I'm sure.*
*Everything I build is mine, and every volt I make is pure.*
*Take away your integration; let us learn and let us teach,*
*Oh, beware this epidemic Berkeleyitis, I beseech.*
*Oh, dammit! Engineering isn't physics, is that plain?*
*Take, oh take, your billion dollars, let's be physicists again.'*

# Ten years later—1956

## A SEQUEL

*Within the halls of NSF the panelists assemble.*
*From all the land the experts band their wisdom to exchange.*
*A great man stands to speak and with applause the rafters tremble.*
*'My friends,' says he, 'we all can see that budgets now must change.*
*By toil and sweat the Soviet have reached ten billion volts.*
*Shall we downtrodden physicists submit? No, no,—revolt!*
*It never shall be said that we let others lead the way.*
*We'll band together all our finest brains and save the day.*

*Give us back our billion dollars, better add ten billion more.*
*If your budget looks unbalanced, just remember this is war.*
*Never mind the Army's shrieking, never mind the Navy's pain.*
*Never mind the Air Force projects disappearing down the drain.*
*In coordinates barycentric, every BeV means lots of cash,*
*There will be no cheap solutions,—neither straight nor synchroclash.*
*If we outbuild the Russians, it will be because we spend.*
*Give, oh give those billion dollars, let them flow without an end.*

[*Folklore records that the brave and solitary scientist who so vigorously*
*defended the purity of science at the original meeting was killed by*
*a beam of hyperons when the Berkeley Bevatron was first switched on.*]

## Standards for inconsequential trivia

PHILIP A SIMPSON

From *The NBS*
*S andard* 15
(1 January 1970).

| | |
|---|---|
| $10^{-15}$ bismol | $= 1$ femto-bismol |
| $10^{-12}$ boos | $= 1$ picoboo |
| $1$ boo$^2$ | $= 1$ boo boo |
| $10^{-18}$ boys | $= 1$ attoboy |
| $10^{12}$ bulls | $= 1$ terabull |
| $10^1$ cards | $= 1$ decacards |
| $10^{-9}$ goats | $= 1$ nanogoat |
| $2$ gorics | $= 1$ paregoric |
| $10^{-3}$ ink machines | $= 1$ millink machine |
| $10^9$ los | $= 1$ gigalo |
| $10^{-1}$ mate | $= 1$ decimate |
| $10^{-2}$ mental | $= 1$ centimental |
| $10^{-2}$ pedes | $= 1$ centipede |
| $10^6$ phones | $= 1$ megaphone |
| $10^{-6}$ phones | $= 1$ microphone |
| $10^{12}$ pins | $= 1$ terapin |

# How radar began

A P ROWE

Extracted from
*One Story of Radar*
(Cambridge
University Press)
1948.

[*In 1934, the problem of getting advance warning against massed air attacks seemed insuperable. The Committee for the Scientific Survey of Air Defence was formed in November of that year, and prepared to consider its first question. At the time, R A Watson-Watt was Superintendent of the Radio Department of the National Physical Laboratory.*]

For many years the 'death ray' had been a hardy annual among optimistic inventors. The usual claim was that by means of a ray emanating from a secret device (known to us in the Air Ministry as a Black Box) the inventor had killed rabbits at short distances and if only he were given time and money, particularly money, he would produce a bigger and better ray which would destroy any object, such as an aircraft, on to which the ray was directed. Inventors were diffident about discussing the contents of their black boxes and, despite the protection afforded by the patent laws, invariably wanted some of the taxpayers' money before there could be any discussion of their ideas. The Ministry solved the problem by offering £1000 to any owner of a Black Box who could demonstrate the killing of a sheep at a range of 100 yards, the secret to remain with its owner.

The mortality rate of sheep was not affected by this offer. The idea of a death ray however was not absurd and something of the kind may come within a hundred years. Because of the recurring claims regarding such rays, there is little doubt that, in writing to Watson-Watt, it was hoped to dispose of the problem, one way or the other, before the Committee met; the problem being whether it was possible to concentrate in an electromagnetic beam sufficient energy to melt the metal structure of an aircraft or incapacitate the crew. Watson-Watt's answer to the death-ray question was a simple one. He said that, although there was no possibility of directing enough energy on to an aircraft to produce a lethal effect at useful distances, it should be possible to locate the plan position of an aircraft by measuring its distances from two points on the ground. The principle was simple enough. Every schoolboy knows that he can measure his distance from a cliff by timing the interval between his shout and the reception of the echo from the cliff. Watson-Watt proposed that a pulse or 'shout' of electromagnetic energy (which travels at about 186 000 miles per second) should be emitted from a point on the ground so as to be incident on an aircraft which, he calculated, would reflect or re-radiate back sufficient energy to enable an 'echo' to be received.

The principle involved was not new. E V Appleton and others had by this method measured distances from electrically charged layers in the atmosphere and had located the positions of thunderstorms by obtaining reflections from electrically charged clouds.

Realizing that calculations were not enough, the committee wanted the earliest possible practical demonstration that Watson-Watt's proposals were worth pursuing; they wanted a demonstration of what scientists call an 'effect'. Watson-Watt therefore proposed that an aircraft should fly in the 50 metre Daventry (continuous-wave) radio beam used for Empire broadcasting and that he should erect simple equipment on the ground to demonstrate whether sufficient energy was reflected from the aircraft to produce an 'effect' with his instruments.

This was done near Daventry on 26 February 1935. Graphic accounts have been written of this demonstration, of how senior officers from the fighting Services went to Daventry on that great day; how for the first time the position of an aircraft was obtained by radar and how success was hailed with congratulations from the distinguished onlookers. In fact, none of these things happened. Though there was not a demonstration of the location of an aircraft, what happened was significant enough. Overnight one of Watson-Watt's assistants, A F Wilkins, had erected equipment in a van near Daventry. All that was hoped of this equipment was that it would show that an aircraft, when in the Daventry beam, would reflect enough of the beam's energy for its presence somewhere in the vicinity to be inferred from observations in the van. This is just what happened on 26 February 1935. So far from the demonstration being witnessed by distinguished officers from the Services, the sole representative from the Service departments was one humble civilian scientific worker.

We were pleased with the demonstration, since reflections from the aircraft were obtained when it was estimated to be about eight miles away, but we knew that we had not seen the location of an aircraft by radio.

[*Pulse equipment soon demonstrated that the system worked at longer ranges. Within a few years radar had grown into a major industry.*]

In his speech at a conference on accelerators (October 1968, Moscow) Academician M A Markov quoted the words of Joliot-Curie: 'The farther the experiment is from theory the closer it is to the Nobel Prize.'

# Building research

R V JONES

From an address to a conference on the design of physics laboratories, University of Lancaster (1969).

When I was concerned with planning my own building in 1958, I did not foresee that within a few years I should be making geophysical instruments; and, although I had tried to check that the site of the new building was as stable as possible, it has not turned out to be as stable as our old building. The instrument on which the deficiency shows up particularly is a tiltmeter which records the local inclination of the Earth's surface relative to the apparent direction of gravity. I was just starting to make this type of instrument as we moved to the new building; and at the old site I quickly found that there was a rhythmic change in the tilt of the Earth's surface, of the order of a few parts in ten million, which followed the tide in the North Sea. The cause is the extra weight of water at high tide, which compresses the sea-bed and thus tilts the land eastward. One could even 'see' the difference between spring and neap tides. These effects are much harder, and often impossible, to observe in the new building, because there are other and larger effects which often obscure them. I have still not located all the trouble, but one of its causes is the Sun, which seems to repel the building when it comes out. The effect is probably due to the fact that the building is situated on a slight rise, and the heat of the Sun causes the ground to expand more on one side of the building than the other, so that the building tilts away from the Sun.

The whole building floats like a ship in a sea of sand; and—also like a ship—it alters its trim as the weight distribution shifts as people move about. Had I not been there at the time I would have been very puzzled by a rhythmic tilt of period about one minute, which at first looked like an unusual effect of a distant earthquake. The building was gently rocking to and fro with an amplitude of a few parts in a hundred million, owing to the shift of weight as a conscientious cleaner moved herself and her floor polisher backwards and forwards progressively along the building.

In partial compensation for these tribulations, I received a letter some months ago:

'At school, for science, our teacher said that Aberdeen slants when the tide comes in. Our class does not believe this but if it is true we would like some information about why this happens. We hope it isn't too much trouble because all of us are very interested in finding out more about it.

Yours sincerely,
Eleanor Wallace.
(Please write back)'

My reply to this letter met with a warm response:
'Thank you very much for sending me the diagrams and explanation on Aberdeen. It was very good of you to take the time.
Yours faithfully,
Eleanor Wallace
× × ×'
Research sometimes has its unforeseen rewards.

---

## Perils of modern living

H P FURTH

From *The New Yorker* 10 November 1956.

A kind of matter directly opposed to the matter known on earth exists somewhere else in the universe, Dr Edward Teller has said . . . . He said there may be anti-stars and anti-galaxies entirely composed of such anti-matter. Teller did not describe the properties of anti-matter except to say there is none of it on earth, and that it would explode on contact with ordinary matter.
San Francisco Chronicle,

*Well up beyond the tropostrata*
*There is a region stark and stellar*
*Where, on a streak of anti-matter,*
*Lived Dr Edward Anti-Teller.*

*Remote from Fusion's origin,*
*He lived unguessed and unawares*
*With all his anti-kith and kin,*
*And kept macassars on his charis.*

*One morning, idling by the sea,*
*He spied a tin of monstrous girth*
*That bore three letters: A. E. C.*
*Out stepped a visitor from Earth.*

*Then, shouting gladly o'er the sands,*
*Met two who in their alien ways*
*Were like as lentils. Their right hands*
*Clasped, and the rest was gamma rays.*

# Predictions and comments

SMITHSONIAN INSTITUTION

I am tired of all this thing called science here. . . . We have spent millions in that sort of thing for the last few years, and it is time it should be stopped.

Senator Simon Cameron (1901)

AIRCRAFT

We hope that Professor Langley will not put his substantial greatness as a scientist in further peril by continuing to waste his time, and the money involved, in further airship experiments. Life is too short, and he is capable of services to humanity incomparably greater than can be expected to result from trying to fly. . . . For students and investigators of the Langley type there are more useful employments.

*New York Times*, December 10, 1903, editorial page

The demonstration that no possible combination of known substances, known forms of machinery and known forms of force, can be united in a practical machine by which man shall fly long distances through the air, seems to the writer as complete as it is possible for the demonstration of any physical fact to be.

Simon Newcomb (1835-1909)

ALTERNATING CURRENT

There is no plea which will justify the use of high-tension and alternating currents, either in a scientific or a commercial sense. They are employed solely to reduce investment in copper wire and real estate.

My personal desire would be to prohibit entirely the use of alternating currents. They are unnecessary as they are dangerous. . . . I can therefore see no justification for the introduction of a system which has no element of permanency and every element of danger to life and property.

I have always consistently opposed high-tension and alternating systems of electric lighting, not only on account of danger, but because of their general unreliability and unsuitability for any general system of distribution.

Thomas A Edison 1889

ROBERT GODDARD'S ROCKET RESEARCH

That Professor Goddard with his 'chair' in Clark College and the countenancing of the Smithsonian Institution does not know the relation of action to reaction, and of the need to have something

better than a vacuum against which to react—to say that would be
absurd. Of course he only seems to lack the knowledge ladled out
daily in high schools. . . .

*New York Times* editorial 1921

I would much prefer to have Goddard interested in real scientific
development than to have him primarily interested in more spectacular achievements which are of less real value.

Charles A Lindbergh to the Guggenheim Foundation 1936

BOMBING FROM AIRPLANES

As far as sinking a ship with a bomb is concerned, you just can't
do it.

US Rear-Admiral Clark Woodward (1939)

POSSIBILITY OF INTERCONTINENTAL MISSILES

There has been a great deal said about a 3000 miles high-angle
rocket. In my opinion such a thing is impossible for many years.
The people who have been writing these things that annoy me,
have been talking about a 3000 mile high-angle rocket shot from
one continent to another, carrying an atomic bomb and so directed
as to be a precise weapon which would land exactly on a certain
target, such as a city.

I say, technically, I don't think anyone in the world knows how
to do such a thing, and I feel confident that it will not be done for
a very long period of time to come. . . . I think we can leave that
out of our thinking. I wish the American public would leave that
out of their thinking.

Dr Vannevar Bush (1945)

THE ATOMIC BOMB

That is the biggest fool thing we have ever done. The bomb will
never go off, and I speak as an expert in explosives.

Adm William Leahy to President Truman 1945

PROPOSAL TO DRIVE A STEAMBOAT BY
SCREW-PROPELLER

Even if the propeller had the power of propelling the boat, it
would be found altogether useless in practice, *because* the power
being applied in the *stern* it would be *absolutely impossible* to make
the vessel steer.

Sir William Symonds, Surveyor of the Royal Navy (1837)

RADIO

In 1913 Lee de Forest, inventor of the audion tube, was brought to trial on charges of fraudulently using the US mails to sell the public stock in the Radio Telephone Company. The District Attorney charged that

'De Forest has said in many newspapers and over his signature that it would be possible to transmit the human voice across the Atlantic before many years. Based on these absurd and deliberately misleading statements, the misguided public . . . has been persuaded to purchase stock in his company.'

*Little Willie,*
Lovingly col-
lected by Dorothy
Rickard, Illus-
trated by Robert
Day (New York:
Doubleday) 1953.

*Little Willie, full of glee,*
*Put radium in Grandma's tea.*
*Now he thinks it quite a lark*
*To see her shining in the dark.*

# Which units of length?

PAMELA ANDERTON

Units of length have been available to the general public for a long time but the recent drive to advertise one particular brand has led us to publish this report for the assistance of our members.

## BRANDS

We found that the units fell into fairly well defined brands or 'systems' from which we have selected three in general use. Two of these, the 'Rule of Thumb' and the 'British' (known as 'Imperial Standard' in the days when we had an empire) are manufactured in this country; the third, the 'Metric', is imported but fairly readily obtainable.

## TESTS

We asked a panel of members to use units of the selected brands and to comment on their convenience. We also submitted samples to a well-known laboratory to find out how reliable they were. The selected units and the results of the tests are listed in the table.

| Brand | Unit | Reliability | Convenience in use |
|---|---|---|---|
| Metric | 'micron' | excellent | fair[1] |
| British | thou | good | good |
| Rule of Thumb | hair's breadth | poor | hopeless |
| Metric | millimetre | excellent | fair[2] |
| British | inch | good | good |
| Rule of Thumb | thumb | poor | excellent |
| Metric | metre | excellent | good |
| British | yard | fair to good[3] | good |
| | foot | good | good |
| Rule of Thumb | pace of stride | fair | excellent |
| | foot (ie size of shoe) | fair to good[4] | excellent |

[1] Difficult to handle for everyday use and available to special order only
[2] Our panel found it about 25·4 times too small
[3] Some samples tended to shrink
[4] Users with big feet get better results

## CONCLUSIONS

The 'Rule of Thumb' was cheap, robust, very convenient and readily obtainable. On the other hand, it was not sufficiently accurate for all purposes.

The 'British' was convenient and readily obtainable, but some doubts exist as to its reliability. Nevertheless, it seems likely to remain popular for a long time.

The 'Metric' is very reliable but not always as convenient to use as the other brands.

BEST BUYS

For general use—Rule of Thumb
For scientists and for others whose arithmetic is weak—Metric.

---

## Alpher, Bethe and Gamow

Condensed from *Reflections on 'Big Bang' Cosmology* by R A Alpher and R Herman, General Electric Research and Development Center Technical Information Series No 69-C-165, May 1969, p 6.

[*In contrast to the bogus paper* 'Remarks on the Quantum Theory of the Absolute Zero of Temperature' (p 24), *the paper* 'The Origin of the Chemical Elements' *published in 1948, was entirely serious. It proposed the neutron-capture theory of formation of the elements. Only the names of the authors had a spurious ring—in fact it is usually referred to as the* $\alpha\beta\gamma$ *paper. The theory was evolved by Alpher under Gamow's direction.*]

Meetings with Gamow during the course of the thesis work were primarily progress reports followed by wide-ranging discussions of physics. Those meetings were usually held in the late afternoon at a dimly lit bar and grill called 'Little Vienna' near the campus of George Washington University. The fare occasionally made for an interesting state for both student and professor at lectures later in the evening.

Once Gamow, with the usual twinkle in his eye, suggested that we add the name of Hans Bethe to an Alpher-Gamow letter to the Editor of the Physical Review, with the remark '*in absentia*' after the name. At some point between receipt of the manuscript at Brookhaven and publication in the April 1, 1948 issue (believe it or not, a date not of our asking), the '*in absentia*' was removed.

Gamow enjoyed the rather considerable publicity it engendered (though his student did not). Watson Davis, then editor of Science Service, wrote a news column on the thesis subject which said in essence that 'the universe had been created' in less than half an hour and more nearly in five minutes—referring in a popular way to the neutron half-life time scale of nucleosynthesis. The response was fascinating, ranging from feature articles, Sunday supplement stories, newspaper cartoons and voluminous mail from religious fundamentalists, to a packed audience of over 200, including members of the press, at the traditionally public (though usually not in this sense) 'defence' of the thesis.

# Electromagnetic units: 1

From *Nature* 130, 987 (1932).

*[Few questions have made physicists lose their sense of humour more often than that debated by a committee in Paris in 1932—electromagnetic units. One topic was: 'Are B and H quantities of the same kind and is their ratio μ a pure numeric? Or should μ be treated as a dimensional quantity?' The committee was divided on national lines, the British pinning their faith on magnetic poles, French physicists favouring the force between two currents as a basis for their system. An unusual note entered into the deliberations at one point.]*

In the course of the discussion, the chairman, Sir Richard Glazebrook, referred to the fact that he was one of the last surviving pupils of Maxwell and he felt convinced from recollections of Maxwell's teaching that he was of the opinion that $B$ and $H$ were quantities of a different kind. When a vote was taken, nine were in favour of treating $B$ and $H$ as quantities of a different nature, whilst three were in favour of regarding $B$ and $H$ as quantities of the same nature.

---

## Electromagnetic units: 2

H B G CASIMIR

From *Helvetica Physica Acta* 41 741 (1968).

*[Although we perhaps pay less regard to the authority of the past, the controversy is by no means dead, because one system of electromagnetic definitions tends to be incorporated into the MKS system of units. The following article appears, incongruously, in a 600-page Festschrift dedicated in 1968 to Georg Busch and published in a volume of Helvetica Physica Acta.]*

Once upon a time there was in a faraway country a great, great kitchen in which many cooks plied their trade and in which there was a great profusion of pots and pans and kettles and cauldrons and bowls and basins of every size and kind and description. Some of these vessels were empty but others contained eggs or rice or apples or spices and many other delectable things. Now the cooks, if they were not busy broiling and baking and cooking and frying and preparing sundry soups and sauces, amused themselves with philosophical speculation and so it came to happen that the art of tagenometry (from ταγηνον, a frying-pan) was developed to great perfection. Sometimes it was even referred to as panmetry, the art of measuring everything, but the ignorant scullions, mis-interpreting the word, promptly also spoke about potmetry, much the same way in which the transatlantic chefs have supple-mented the hamburger with a cheeseburger.

To every vessel tagenometry assigned a volume $V$. This was measured in cubic inches and determined by measuring dimen-sions with great precision and by then applying the formulae of solid geometry or in case of irregular shapes by numerical inte-gration on a beanheaded abacus. But to every vessel there was also assigned an entirely different quantity, the volumetric dis-placement $W$. This was measured in gallons and determined by filling the vessel with water, pouring out the water, weighing said water in pounds avoirdupois, correcting for temperature and dividing by 10. The ratio of volumetric displacement and volume was referred to as the volumetric constant, $\epsilon = W/V$. In the course of time it became clear that this volumetric constant had the same value for every empty vessel; this became known as the volumetric constant of empty space, $\epsilon_0$. But for other vessels the volumetric constant behaved often in an erratic way. It changed after thermal treatment, or simply with time; it depended on the speed of measurement. Also the dynamic behaviour of moving non-empty pans posed curious problems.

One day a wise man entered the kitchen and after having lis-tened to the worried cooks he said: 'I can solve your problems. There is really only one tagenometric quantity, let us call it the volume and measure it in cubic centimetres. Weighing water will give the same value for an empty vessel if you take the weight in grams. So your volumetric constant of empty space is just unity. But in a non-empty pan part of the volume is occupied by edibles like potatoes or pears or plums; let us call this volume $P$. Then, with the water-method you determine $V - P$. In many cases $P$ will be proportional to $V$, that is $P = \kappa V$. Then the water-weight volume, your volumetric displacement, is $W = V - \kappa V = (1 - \kappa)V$ and hence $\epsilon = 1 - \kappa$. What you really should study is $P$ and its dependence on the constitution and preparation of the victuals. And instead of studying the dynamics of a non-empty pan, you should study the motion of the things it contains'.

The cooks understood, yet they looked crestfallen. 'But our beautiful units' they said. 'What about our goldplated pounds and ounces and drams? Look at that wonderful half-perch in yon corner, neatly subdivided into 99 inches. It would be ill-con-venient to change all that'. The wise man smiled. 'There is no real need to change' he said. 'As long as you are sure to remember that $\epsilon_0$ is just a way to change from one unit to another and that $P$ and $\kappa$ are the only physically relevant quantities, you can work in any system of units you like'.

The years went by. The wise man had died, new generations of cooks worked in the kitchen and got restive over the principles of tagenometry. 'How crazy', they said. 'Isn't it obvious that $V$ and $W$ are quite different quantities, since they are determined in quite different ways? And why should the volumetric constant of empty space be unity? Is a pot of rice not just as good or better h an an empty pot?' These protests prevailed. It was decided at an international congress that even if volume and volumetric displacement were identical in magnitude the one should be measured in Euclid—this being a cubic centimetre—the other in Archimedes. The volumetric displacement of empty space— although equal to unity—had the dimension Archimedes/Euclid. And after having created order in this way, the new generation has returned to inches and pounds, and brands as reactionary anyone who heeds the wise lessons of the wise man.

That is how today's cooks spend their moments of leisure; let us hope that their cuisine will not suffer.

## British units

One recommended British unit of thermal conductivity—useful for calculating the heat transmission of walls—is:

$$\text{BThU/hour/sq ft/cm/}°\text{F}$$

### UNIT OF CAPACITY

From the *Admiralty Handbook of Wireless Telegraphy* 1931.

The *jar* is the Service unit, and is very useful when dealing with the small capacities met with in ordinary wireless practice.

1 farad $= 9 \times 10^8$ (nine hundred million) jars
$1\,\mu\text{F}$ $= 900$ jars

[*It was a big jar, 10 metres in radius, more like a balloon. By 1937 this unit was made obsolete or at any rate obsolescent to bring Royal Naval practice into line with commercial.*]

# Therapy

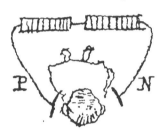

## J P JOULE

From a notebook in the custody of the Director of the Manchester Museum of Science & Technology.

[*James Prescott Joule was already well known at the age of 23 for his experiments on the design and efficiency of electric motors and for his enunciation of the $i^2R$ law. But galvanism remained mysterious. In 1840 he tried the effect of electricity on a lame horse and the next year he recorded the following. Who 'she' was and whether the word 'patient' meant a sick or mentally sick person, cannot be determined.*]

1841, May 31. Monday 6.00 PM. I took two batteries, each consisting of 10 double pairs (copper & zinc) charged with very dilute S. acid, and connected their extremeties with copper plates (4 ins square) by means of copper wire; between these plates and the skin of the patient two pieces of flannel soaked in salt water were placed. Negative on the right cheek. Plates half way between the chin and the ears. The action was continued for about 13 minutes during which she felt the usual pricking sensation, with the tremulous feeling all over the face and neck, terminating at the shoulders and eyes, and occasioning a strong taste in the mouth. The flannel and plates were then removed so as to cover the ears for one minute when she felt a very strong action through the head, her eyes shut and she quivered very strongly, and she fainted, and it was thought advisable to terminate the operations.

---

Newspaper report.

Air sacs give birds buoyancy in flight. To watch a large bird on a quiet summer's day keep or gain altitude in spiraling, soaring flight on uprising air currents of such slight lifting power that they will scarcely support a dust particle or a tiny-winged insect, makes the beholder know that the bird's buoyancy must be achieved by more than just outward design. And that it is! Scattered within the bird's cavity are at least five air sacs that take up every bit of space not occupied by other organs . . . Besides, these extensive buoyant compartments connect yet farther with many of the bird's hollow bones . . .

# Infancy of x-rays

G E M JAUNCEY

From *American Journal of Physics* 13 December 1945. X-rays were discovered by Wilhelm Conrad Roentgen at Würzburg, Germany, on November 8, 1895. The discovery amazed and excited both physicists and the general public. Newspapers reported wild rumours, extravagant claims, and fanciful speculations.

'It is suggested that, if all that has reached us by cable is true, there will no longer be any privacy in a man's home, as anyone with a vacuum tube outfit can obtain a full view of any interior through a brick wall.'

Another news item suggested that an x-ray could bring back life and that cathode rays (there was confusion of cathode rays with x-rays amongst both the general public and the physicists) could be used for resuscitating electrocuted persons.

The possibility of photographing the human skeleton through the flesh amazed the public. The following is quoted from the *Scientific American* of February 22, 1896:

'The new photography has moved the English heart to poetry. The following verses are not by the new Poet Laureate, but they shed new light upon the future uses to which the shadow photographs may be put. Our thanks are due London *Punch*, to whom we are indebted:

*O Roentgen, then the news is true*
    *And not a trick of idle rumour*
*That bids us each beware of you*
    *And of your grim and graveyard humour.*

*We do not want, like Dr Swift,*
    *To take our flesh off and to pose in*
*Our bones, or show each little rift*
    *And joint for you to poke your nose in.*

*We only crave to contemplate*
    *Each other's usual full dress photo;*
*Your worse than 'altogether' state*
    *Of portraiture we bar in toto!*

*The fondest swain would scarcely prize*
    *A picture of his lady's framework;*
*To gaze on this with yearning eyes*
    *Would probably be rated tame work.*

*No, keep them for your epitaph*
    *These tombstone souvenirs unpleasant;*
*Or go away and photograph*
    *Mahatmas, spooks, and Mrs Besant.*

The Mrs Besant of the poem was a prominent English theosophist (some people called her a spiritualist) of 1896.

Finally, Sir Arthur Schuster's description in *The Progress of Physics* of his experiences with the public in early 1896 is worth recording:

'My laboratory was inundated by medical men bringing patients, who were suspected of having needles in various parts of their bodies, and during one week I had to give the best part of three mornings to locating a needle in the foot of a ballet dancer, whose ailment had been diagnosed as bone disease. The discharge tubes had all to be prepared in the laboratory itself, and, where a few seconds exposure is required now [1911], half an hour had to be sacrificed owing to our ignorance of the best conditions for producing the rays.'

Schuster states that such interruptions seriously interfered with his experiments on the magnetic deflection of cathode rays. Schuster's experiences were duplicated in many physics laboratories all over the world.

## Faraday lectures

[*Michael Faraday was a brilliant lecturer: his popular discourses at the Royal Institution were an important means of disseminating scientific knowledge from 1812 onwards. He studied the art of lecturing carefully; here are some of his aphorisms.*]

One hour is long enough for anyone.

Listeners expect reason and sense, whilst gazers only require a succession of words.

The most prominent requisite of a lecturer, though perhaps not the most important, is a good delivery.

The lecturer should give the audience full reason to believe that all his powers have been exerted for their pleasure and instruction.

[*He was highly self-critical of his own abilities. To a friend he wrote:*]

As when on some secluded branch in forest far and wide sits perched an owl, who, full of self-conceit and self-created wisdom, explains, comments, condemns, ordains and orders things not understood, yet full of his importance still holds forth to stocks and stones around—so sits and scribbles Mike.

# N rays

R  W  WOOD

Condensed from
Dr Wood,
Modern Wizard of
the Laboratory by
William Seabrook
(Harcourt Brace)
1941.

[*Here is Wood's own account of what was probably the greatest scientific delusion of our time.*]

In the late autumn of 1903, Professor R Blondlot, head of the Department of Physics at the University of Nancy, member of the French Academy, and widely known as an investigator, announced the discovery of a new ray, which he called N ray, with properties far transcending those of the x-rays. Reading of his remarkable experiments, I attempted to repeat his observations, but failed to confirm them after wasting a whole morning. According to Blondlot, the rays were given off spontaneously by many metals. A piece of paper, very feebly illuminated, could be used as a detector, for, wonder of wonders, when the N rays fell upon the eye they increased its ability to see objects in a nearly dark room.

Fuel was added by a score of other investigators. Twelve papers had appeared in the *Comptes rendus* before the year was out. A Charpentier, famous for his fantastic experiments on hypnotism, claimed that N rays were given off by muscle, nerves, and the brain, and his incredible claims were published in the *Comptes*, sponsored by the great d'Arsonval, France's foremost authority on electricity and magnetism.

Blondlot next announced that he had constructed a spectroscope with aluminium lenses and a prism of the same metal, and found a spectrum of lines separated by dark intervals, showing that there were N rays of different refrangibility and wave length. He measured the wave lengths. Jean Becquerel claimed that N rays could be transmitted over a wire. By early summer Blondlot had published twenty papers, Charpentier twenty, and J Becquerel ten, all describing new properties and sources of the rays.

Scientists in all other countries were frankly skeptical, but the French Academy stamped Blondlot's work with its approval by awarding him the Lalande prize of 20000 francs and its gold medal 'for the discovery of the N rays.'

In September (1904) I went to Cambridge for the meeting of the British Association for the Advancement of Science. After the meeting some of us got together for a discussion of what was to be done about the N rays. Professor Rubens, of Berlin, was most outspoken in his denunciation. He felt particularly aggrieved because the Kaiser had commanded him to come to Potsdam and demonstrate the rays. After wasting two weeks in vain attempts to duplicate the Frenchman's experiments, he was greatly embarrassed by having to confess to the Kaiser his failure. Turning to me he said, 'Professor Wood, will you not go to Nancy immediately and test the experiments that are going on there?' 'Yes, yes,' said all of the Englishmen, 'that's a good idea, go ahead.' I suggested that Rubens go, as he was the chief victim, but he said that Blondlot had been most polite in answering his many letters asking for more detailed information, and it would not look well if he undertook to expose him. 'Besides,' he added, 'you are an American, and you Americans can do *anything*. . . '.

So I visited Nancy, meeting Blondlot by appointment at his laboratory in the early evening. He spoke no English, and I elected German as our means of communication, as I wanted him to feel free to speak confidentially to his assistant.

He first showed me a card on which some circles had been painted in luminous paint. He turned down the gas light and called my attention to their increased luminosity, when the N ray was turned on. I said I saw no change. He said that was because my eyes were not sensitive enough, so that proved nothing. I asked him if I could move an opaque lead screen in and out of the path of the rays while he called out the fluctuations of the screen. He was almost 100 per cent wrong and called out fluctuations when I made no movement at all, and that proved a lot, but I held my tongue. He then showed me the dimly lighted clock, and tried to convince me that he could see the hands when he held a large flat file just above his eyes. I asked if I could hold the file, for I had noticed a flat wooden ruler on his desk, and remembered that wood was one of the few substances that *never* emitted N rays. He agreed to this, and I felt around in the dark for the ruler and held it in front of his face. Oh, yes, he could see the hands perfectly. This also proved something.

But the crucial and most exciting test was now to come. Accompanied by the assistant, who was by this time casting rather hostile glances at me, we went into the room where the spectroscope with the aluminium lenses and prism was installed. In place of an eyepiece, this instrument had a vertical thread, painted with luminous paint, which could be moved along in the region where the N ray spectrum was supposed to be by turning a wheel having graduations and numerals on its rim. Blondlot took a seat in front of the instrument and slowly turned the wheel. The thread was supposed to brighten as it crossed the invisible lines of the N-ray spectrum. He read off the numbers on the graduated scale for a number of the lines, by the light of a small, darkroom red lantern. This experiment had convinced a number of skeptical visitors, as he could repeat his measurements in their presence, always getting the same numbers. I asked him to repeat his measurements, and reached over in the dark and lifted the aluminium prism from the spectroscope. He turned the wheel again, reading off the same numbers as before. I put the prism back before the lights were turned up, and Blondlot told his assistant that his eyes were tired. The assistant had evidently become suspicious, and asked Blondlot to let him repeat the reading for me. Before he turned down the light I had noticed that he placed the prism very exactly on its little round support, with two of its corners exactly on the rim of the metal disk. As soon as the light was lowered, I moved over towards the prism, with audible footsteps, but *I did not touch the prism*. The assistant commenced to turn the wheel, and suddenly said hurriedly to Blondlot in French, 'I see nothing; there is no spectrum. I think the American has made some *dérangement*.' Whereupon he immediately turned up the gas and went over and examined the prism carefully. He glared at me, but I gave no indication of my reactions. This ended the seance.

Next morning I sent off a letter to *Nature* giving a full account of my findings, not, however, mentioning the double-crossing incident at the end of the evening, and merely locating the laboratory as 'one in which most of the N-ray experiments had been carried out.' *La Revue scientifique*, France's weekly semipopular scientific journal started an inquiry, asking French scientists to express their opinions as to the reality of the N rays. About forty letters were published, only a half dozen backing Blondlot. The most scathing one by Le Bel said, 'What a spectacle for French science when one of its distinguished savants measures the position of the spectrum lines, while the prism reposes in the pocket of his American colleague!'

The Academy at its annual meeting in December, when the prize and medal were presented, announced the award as given to Blondlot 'for his life's work, taken as a whole.'

[*Seabrook adds: The tragic exposure eventually led to Blondlot's madness and death. He was a great man, utterly sincere, who had 'gone off the deep end,' perhaps through some form of self-hypnotism. . . . What Wood had done, reluctantly but with scientific ruthlessness, had been the coup de grace.*]

# My initiation

L  ROSENFELD

Condensed from
*Journal of Jocular
Physics* vol 2 p 7
(October 1945)
Institute of
Theoretical
Physics,
Copenhagen.

The first message I got from Bohr was a telegram, announcing that the Easter Conference was to be postponed two days. I was then —in 1929—in Göttingen and, together with Heitler, had expressed the wish to attend that famous Conference; we had both received from Klein a favourable answer, to which the aforementioned telegram brought the master's eleventh-hour correction. When we arrived in Copenhagen, Bohr informed us of the reason for the postponement: he had to complete ('with Klein's kind help') a Danish translation of some of his recent papers to be published as a Festskrift of Copenhagen University; he told us all about this venerable Festskrift tradition and added: 'It would have been a catastrophe if that work had not been ready in time!' This struck me as a hyperbolic way of stating the matter. How little I imagined at that moment the tragedy hidden behind this seemingly innocuous procedure of putting the finishing touch to a paper! How little I knew that it was my destiny to play a part in a whole lot of such tragedies!

My sole excuse for the failure to grasp the earnestness of this paper-writing subject is that I was by no means an exception in that respect. In fact, as experience taught me since, people are on the whole distressingly unimaginative on that point. Take, for instance, the case of the Faraday Lecture. Bohr arrived in London for the Faraday celebration with a manuscript of his lecture, which he described as 'practically finished.' There were just a few pages lacking. The plan was to seek the required isolation in the romantic environment of some old English inn, and in a week's time, 'with Rosenfeld's kind help', (he explained to Mr Carr, the secretary of the Chemical Society) the thing would be definitively disposed of. Mr Carr was delighted. After a week's hard labour in a rather crowded and thoroughly unromantic hotel, in which we had to wage a regular war of nerves against an irascible schoolmistress for the possession of the parlour, the ten odd lacking pages had actually been written. We had furthermore gained the insight that a great improvement could be obtained by the mere addition of some twenty more pages. Bohr quite warmed up at this idea, which (he persuaded me) brought us really a good deal nearer to the end. I was accordingly dispatched to Mr Carr to report on the new development. Well, Mr Carr did not at all cheer the prospect; he was just annoyed; he even made no effort to conceal his annoyance. When I alluded to our having worked the whole week without respite, I am sorry to say he looked decidedly incredulous. I was quite downhearted when I left him. Fortunately, I had just then an appointment with Delbrück, whom I found in company of one of his innumerable lady-friends. *He* was a man of feeling and understanding; he comforted me like the true friend he was.

But to return to the scene on the platform in Copenhagen station. What impressed me most about Bohr at this first meeting, was the benevolence radiating from his whole being. There was a paternal air about him, which was enhanced by the presence of some of his sons. These sons of Bohr's were a great matter of speculation to me. When I again saw Bohr at the Institute the next morning, there were also a few sons around him. Different ones, I thought; he must have a host of them. On the afternoon of the same day, however, I was bewildered at the sight of still another son at his side. He seemed to stamp them from the ground or draw them forth from his sleeve, like a conjurer. At length, however, I learned to distinguish the sons from one another and I found out that there was only a finite number of them after all.

I don't know how the Athenian delegates for oracle consultation felt on their return from Delos. But I imagine their feelings must have been akin to mine after I had listened to Bohr's introductory lecture at the Conference. He had begun with a few general considerations calculated, no doubt, to convey to the audience that peculiar sensation of having the ground suddenly removed from under their feet, which is so effective in promoting receptiveness for complementary thinking. This preliminary result being readily achieved, he had eagerly hastened to his main subject and stunned us all (except Pauli) with the non-observability of the electron spin. I spent the afternoon with Heitler pondering on the scanty fragments of the hidden wisdom which we had been able to jot down in our note books. Towards the evening we felt the need for some fortification and proceeded to the Strøg.

The following evening we spent at the cinema, together with some others. Picture theatres have always been institutions of high educational value to young theoretical physicists. So it turned out this time too. There it was that Casimir started his important calculation of the magnetic field exerted by an atomic Dirac electron on the nucleus of the atom. He had to work under very trying circumstances. For as soon as any part of the show started, the lights went out, and poor Casimir had to wait until the lovers had safely got over their troubles and married and all before he could resume his calculations. He did not lose a second either: every time the lights went up, they invariably disclosed our friend bent over odd bits of paper and feverishly filling them with intricate formulae. The way in which he made the best of a desperate situation was truly admirable. It was inspiring to watch him.

On the last day of the Conference I experienced the climax of my Copenhagen initiation. It came about rather unexpectedly in the following way. At the meeting that morning one of the most distinguished guests had developed some views about the vexed question of the 'cut' between system and observer, which seemed to me rather erroneous. Bohr, however, had only opposed them feebly (as I thought); in his rather confused speech, the phrase 'very interesting' recurred insistently; and finally, turning to the speaker, he had concluded by expressing the conviction that 'we agree much more than you think.' I was much worried by this extraordinary attitude, the more so as the highbrow bench seemed to find it all right. I therefore ventured to explain my doubts straight away to Bohr. I began cautiously to state that the speaker's argument did not seem to me quite justified. 'Oh,' said Bohr quickly, 'it is pure nonsense!' So I knew I had been led astray by a mere matter of terminology.

But now the unexpected happened. Bohr summoned me to a little room, in the middle of which stood a rather long table. He manoeuvred me towards that table and as soon as I stood leaning against it, he began to describe around it, at a rather lively pace, a keplerian ellipse of large eccentricity, of which the place where I was standing was a focus. All the time, he was talking in a soft low voice, explaining to me the broad outlines of his philosophy. He walked with bent head and knit brows; from time to time, he looked up at me and underlined some important point by a sober gesture. As he spoke, the words and sentences which I had read before in his papers suddenly took life and became loaded with meaning. It was one of the few solemn moments that count in an existence, the revelation of a world of dazzling thought, truly an initiation.

It is generally recognized that no initiation can be properly accomplished without being combined with a painful experience of some sort. In that particular also my initiation left nothing to be desired. For since I had to strain my hearing to the utmost to catch the master's words, I was compelled to execute a continuous rotation at the same rate as that of his orbital motion. The true purpose of the ceremony, however, did not occur to me until

Bohr ended by emphasizing that you can't even catch a glimpse of complementarity if you don't feel completely dizzy. When I heard that, I realized everything and I could only pay him a silent homage of thankfulness and admiration for such touching solicitude.

## Frank Jewett

PAUL E KLOPSTEG

From *Science* 140 pp 594–8 (1963).

Frank Jewett, late president and chairman of Bell Telephone Laboratories, was president of the National Academy of Sciences during World War II. One evening at the old Cosmos Club, he regaled a small group of us with an autobiographical note which had a flavour of biophysics.

During his childhood in Pasadena he and some of his friends, aged about ten, became interested not only in bird-watching, but also in studying the habits and life histories of birds. Jewett chose hummingbirds for his study, many species of which congregated in the Pasadena area during the winter months. With the onset of migration, each species sorted itself out from the others and took off for its summer habitat. One fact of particular interest to him was the cleanliness of the hummingbird's nest as compared with that of any other kind of bird. Careful observation disclosed the reason. The earliest training given a chick by the mother hummingbird was 'toilet training' of sorts. It consisted of teaching the chick, immediately upon emergence from its shell, to elevate its posterior above the edge of the nest when defecating.

Jewett's interest in physics suggested the possibility of a simple experiment based on this observation. He measured the height of the nest above the ground, and the horizontal distance of the droppings from the vertical to the nest. These data enabled him to determine the initial velocity, assuming horizontal propulsion. It proved astounding that a hummingbird chick, weighing only a few grams, could muster such propulsive energy. Jewett speculated on the validity of extrapolating from the velocity–weight relation for a few grams of body weight to velocities for greater body weights, say up to 75 kilograms. The reader's imagination can readily supply the discussion about these speculations.

# Inertia of a broomstick

From *Popular Scientific Recreations in Natural Philosophy* by Gaston Tissandier, (London: Ward Lock) 1881.

[*The splendid gentleman in the picture opposite is performing an experiment popular a century ago. Instructions are:*]

A needle is fixed at each end of the broomstick, and these needles are made to rest on two glasses, placed on chairs; the needles alone must be in contact with the glasses. If the broomstick is then struck violently with another stout stick, the former will be broken, but the glasses will remain intact. The experiment answers all the better the more energetic the action. It is explained by the resistance of inertia in the broomstick. The shock suddenly given, the impulse has not time to pass on from the particles directly affected to the adjacent particles; the former separate before the movement can be transmitted to the glasses serving as supports.

I believe that I am not overstating the truth when I say that half the time occupied by clerks and draftsmen in engineers' and surveyors' offices . . . is work entailed upon them by the present farrago of weights and measures.

*Lord Kelvin*

Letter to *Physics Today*, November 1964

Once again (F Bulos *et al*, *Phys. Rev. Letters* **13**, 486 (1964)) the high energy physicists have presented us with a paper that has more authors (27) than paragraphs (12). Can high energy really be so different?

*Robert A Myers*

*Oh Langley devised the bolometer:*
*It's really a kind of thermometer*
*Which measures the heat*
*From a polar bear's feet*
*At a distance of half a kilometre.*

See the stars at home with Spitz planetarium, $14.95 . . . reflects all the major constellations of the Western Hemisphere on the ceiling of a darkened room . . .

(Advertisement)

*Scientific Researches! — New Discoveries in PNEUMATICKS! — or — an Experimental Lecture — Powers of Air*

# Pneumatic experiment

Cartoon by
James Gillray,
1802.

*[Experiments to determine the effects of inhaling different gases* **were**
*important in the early years of the nineteenth century as new gases were
discovered. On March 22, 1800 Lady Holland went to the Royal
Institution for a lecture–demonstration. She wrote:]*

This Institution of Rumford's furnishes ridiculous stories. The
other day they tried the effect of the gas, so poetically described
by Beddoes; it exhilarates the spirits, distends the machine. The
first subject was a corpulent middle-aged gentleman, who, after
inhaling a sufficient dose, was requested to describe to the company his sensations; 'Why, I only feel stupid'. This intelligence
was received amidst a burst of applause, most probably not for the
novelty of the information. Sir Coxe Hippisley was the next who
submitted to the operation, but the effect upon him was so animat-
ing that the ladies tittered, held up their hands, and declared themselves satisfied.

*[The demonstrator administering the nitrous oxide is probably Thomas
Young, professor of natural philosophy and chemistry, with his assistant
Humphry Davy at his side. Rumford stands on the right, Isaac D'Israeli
sits at the far right. Hippisley and Rumford were among the founders of
the Royal Institution. Note the smouldering candle and tobacco pipe ready
for igniting by being put in the jar of oxygen.]*

## The high standard of education in Scotland

SIR W L BRAGG

We were staying in Ballater, a small town on Deeside in Scotland.
In the town was a tiny shop which sold tourist attractions and
picture postcards, and in its minute window was a very fine
specimen of smoky quartz mineral. Buying a postcard, I said to
the proprietor, 'That's a fine group of smoky quartz in your
window' and had this reply in very broad Scotch:
'That's no smoky quartz, that's topaz. It's a crystal. You can
tell crystals by the angles between their faces. If you're interested
I'll lend you a book on the subject.'
I knew enough (crystals being rather in my line) to be sure it
was smoky quartz, and on return to base looked up a book on
Mineralogy which said 'Smoky Quartz, also known as Cairngorm,
is called Topaz in Scotland'.

# Theoretical zipperdynamics

HARRY J ZIPKIN *Department of Unclear Phyzipics,
The Weizipmann Inziptute*

From *Journal of
Irreproducible
Results*, 3, 6 (1956).

## INTRODUCTION

The fundamental principles of zipper operation were never well understood before the discovery of the quantum theory [1]. Now that the role of quantum effects in zippers has been convincingly demonstrated [2], it can be concluded that the present state of our knowledge of zipper operation is approximately equal to zero. Note that because of the quantum nature of the problem, one cannot say that the present state of knowledge is *exactly* equal to zero. There exist certain typically quantum-mechanical zero-point fluctuations; thus our understanding of the zipper can vary from time to time. The root mean square average of our understanding, however, remains of the order of *h*.

### ZIPPERBEWEGUNG

The problem which baffled all the classical investigators was that of 'zipperbewegung' [3], or how a zipper moves from one position to the next. It was only after the principle of complementarity was applied by Niels Bohr [4], that the essentially quantum-theoretical nature of the problem was realized. Bohr showed that each zipper position represented a quantum state, and that the motion of the zipper from one position to the next was a quantum jump which could not be described in classical terms, and whose details could never be determined by experiment. The zipper just jumps from one state to the next, and it is meaningless to ask how it does this. One can only make statistical predictions of zipperbewegung.

The unobservability of zipperbewegung is due, as in most quantum-phenomena, to the impossibility of elimination of the interaction between the observer and the apparatus. This was seriously questioned by Einstein who, in a celebrated controversy with Bohr, proposed a series of experiments to observe zipperbewegung. Bohr was proved correct in all cases; in any attempt to examine a zipper carefully, the interaction with the observer was so strong that the zipper was completely incapacitated [5].

### THE SEMI-INFINITE ZIPPER

A zipper is a quantum-mechanical system having a series of equally spaced levels or states. Although most zippers in actual use have only a finite number of states, the semi-infinite zipper is of considerable theoretical interest, since it is more easily treated theoretically than is the finite case. This was first done by Schroedzipper [6] who pointed out that the semi-infinite series of equally spaced levels was also found in the Harmonic Oscillator discovered by Talmi [7]. Schroedzipper transformed the zipper problem to the oscillator case by use of a Folded – Woodhouse Canonical Transformation. He was then able to calculate transition probabilities, level spacings, branching ratios, seniorities, juniorities, *etc.* Extensive tables of the associated Racah coefficients have recently been computed by Rose, Bead and Horn [8].

Numerous attempts to verify this theory by experiment have been undertaken, but all have been unsuccessful. The reason for the inevitability of such failure has been recently proved in the celebrated Weisgal–Eshkol theorem [9], which shows that the construction of a semi-infinite zipper requires a semi-infinite budget, and that this is out of the question even at the Weizipmann Inziptute.

Attempts to extend the treatment of the semi-infinite zipper to the finite case have all failed, since the difference between a finite and a semi-infinite zipper is infinite, and cannot be treated as a small perturbation. However, as in other cases, this has not prevented the publishing of a large number of papers giving perturbation results to the first order (no one publishes the higher order calculations since they all diverge). Following the success of M G Mayer [10] who added spin–orbit coupling to the harmonic oscillator, the same was tried for the zipper, but has failed completely. This illustrates the fundamental difference between zippers and nuclei and indicates that there is little hope for the exploitation of zipperic energy to produce useful power. There are, however, great hopes for the exploitation of zipperic energy to produce useless research.

THE FINITE ZIPPER

The problem of the finite zipper is best treated directly, without reference to the infinite case. One must first write the Schroedzipper equation for the system:

$$H(Z) = -i\hbar \, dZ/dt.$$

The solution of this equation is left as an exercise for the reader. From the result all desired observable information can be calculated.

The most interesting case of the finite zipper is that in which there are perturbations. For this case the Schroedzipper equation becomes:

$$(H + H') Z = -i\hbar \, dZ/dt.$$

Because of the perturbation term $H'$, the original states of the unperturbed zipper are no longer eigenstates of the system. The new eigenstates, characteristic of a perturbed zipper, are mixtures of the unperturbed states. This means, roughly, that because of the perturbation the zipper is in a state somewhere in between its ordinary states.

A theoretical possibility of such perturbation was recently voiced by a lady who was considering buying a pair of trousers for her husband. She was offered a zippered type but declined the offer. Her uncertainty principle was expressed in the following words: 'I don't think such trousers would be good for my husband. Last time I bought him a zippered sweater, his tie was highly disturbed by the zipper perturbation'.

REFERENZIPS

1 H Quantum '*A New Theory of Zipper Operation which is also incidentally applicable to such minor Problems as Black Body Radiation, Atomic Spectroscopy, Chemical Binding and Liquid Helium*'. *ZIP* **7**, 432 (1922)

2 H Eisenzip '*The Uncertainty Principle in Zipper Operation*', *Zipschrift für Phyzip*, **2**, 54 (1923)

3 I Newton, M Faraday, C Maxwell, L Euler, L Rayleigh, and J W Gibbs, '*Die Zipperbewegung*' (unpublished)

4 N Bohr '*Lecture on Complementarity in Zippers*', Geneva Conference, '*Zippers for Peace*' (1924)

5 P R Zipsel and N Bohm *Einstein Memorial Lecture*. Haifa Technion (1956)

6 F Schroedzipper '*What is a Zipper*', Dublin (1950)

7 E Talmi, *Helv. Phys. Acta*, **1**, 1 (1901)

8 M E Rose, A Bead, and Sh Horn (to be published)

9 M Weisgal and L Eshkol 'Zippeconomics' *Ann. Rept. Weizipmann Inziptute* (1955)

10 Metro G Mayer '*Enrichment by the Monte Carlo Method: Rotational States with Magic Numbers*', *Gamblionics*, **3**, 56 (1956)

In a summary of lectures on electrodynamics delivered at Moscow University by A A Blasov the following sentence occurred: 'The purpose of the present course is the deepening and development of difficulties underlying contemporary theory . . .'.

California constantly emits neutrons, which strike other materials and make them radioactive—*Birmingham (Ala) News*

[*And it does it in the most blatant sort of way.*]

# Atomic medicine

JOHN H LAWRENCE

From *California Monthly* December 1957 p 17–21.

It can be said that the new field of atomic medicine actually began at the University of California, where artificial radioactivity first became available for biological and medical research. Watching all the young men working around the cyclotron bombarding new targets and measuring the radiations with Geiger counters and Wilson cloud chambers, I was soon infected with the excitement of the early experiments. Very little was known of the biological effects of the neutron rays produced by the cyclotron, and this seemed an important place to start work.

For the neutron ray exposures in Berkeley we made a small metal cylinder to house a rat so that it could be placed close to the cyclotron. After placing the rat in position, we asked the crew to start the cyclotron and then turn it off again after the first two minutes. This 'two-minute' exposure was arbitrary, since we had no basis for calculating how great a dose would produce an observable radiation effect on the animal. After the two minutes had passed, we crawled into the small space between the dees of the 37-inch cyclotron, opened the cylinder, and found the rat was dead. Everyone crowded around to look at the rat, and a healthy respect for nuclear radiations was born. Now, of course, radiation protection measures are an integral part of all atomic energy research programs, but I think this incident of our first rat played a large part in the excellent safety record at the University. In fact, we have had no radiation cataracts among the early cyclotron workers. We discovered later that the rat's death had resulted from asphyxiation rather than radiation. But since our failure to aerate the rat chamber adequately had brought about such a salutory effect on the crew, the post-mortem report was not widely circulated.

The physicists were so busy and excited about their work they did not like to allow us exposure time for the animal experiments and thought us nuisances. One day as I walked by the cyclotron, a pair of pliers thoughtlessly left in my pocket were torn free by the intense magnetic field and flew into the dees of the vacuum chamber of the cyclotron, putting it out of operation for three days. We were even less popular after this incident.

Truth comes out of error more easily than out of confusion.

FRANCIS BACON

# 100 authors against Einstein

From *Die Naturwissen-schaften* 11, 254-6 (1931).

[*A quarter of a century after Einstein published his work on Relativity, a book was published in Germany called '100 Authors against Einstein' which sought to show that Einstein must be wrong because so many opinions were ranged against him. The book was reviewed by von Brunn, and his article is itself an entertaining polemic. Here are some extracts.*]

It is not really surprising that many people who have experienced the development of the theory of Relativity only in its most superficial forms have received an adverse impression of it. For without any fault on the part of the creator of the theory, major tactical errors have been committed by overzealous but not uniformly comprehending proponents of it. The experiments of leaving the verdict on the theories to the *vox populi* have, thank God, been eventually checked by sensible advocates of the new concept. And the tactic practised by certain fanatical scientific supporters of Einstein's theory, of cutting discussion of it short by threatening to discredit even the most moderate and reasonable criticism, as obviously resulting from stupidity and malice—this too has by now been abandoned. But even apart from these excesses of the 'Einstein craze' which now are a thing of the past, serious and respectable grounds for certain misgivings with regard to Relativity Theory do still remain. Even Special Relativity Theory demands certain 'sacrifices of intellect'—in particular the relinquishment of the strict determinability of simultaneity (note that the concept of simultaneity still remains). For many philosophers this is of course tantamount to an irreparable crime against the eternal infallibility of Kant, because they do not understand the inevitability of Einstein's intentions.

If the book under review had arisen solely from fears, justified in principle although very exaggerated, of an 'evaporation of the concept of reality' in modern science, then one would be prepared to tolerate it. But an author who, without sufficient independent personal judgement, collects other people's criticisms of a scientific theory for a tendentious purpose must, quite apart from the moral appraisement of his aim, accept the consequences and his action be dismissed as pamphleteering. But even the most tolerant critic will not be able to find any extenuating circumstances for this 'book by a hundred authors', for what does this council of judges consist of? Ninety per cent of the authors are dyed-in-the-wool Kantians who have not a clue about the crisis of modern physicists with regard to the theory of cognition, a crisis brought about by the failure of all attempts to prove absolute motion using optical means and by the proportionality of inertial and gravitational mass. Their rantings and ravings therefore carry no weight. What does one do—to quote but a few statements which at least make sense—with such pearls of wisdom as:

Einsteinism maintains the equivalence of acceleration and gravitation. In other words, it preaches that an effect (acceleration) is equivalent to its cause (gravitation). This thesis is a blatant absurdity (*Dr A Reuterdahl*).

Or:

The theory of Einstein is for me a functional deformation of reality. His framework of reference, variable space and time coordinates, invariant velocity of light (in spite of variable limiting value), is not to my taste (*Professor Strehl*).

And those are not the worst examples. It is impossible in a review to examine these 'arguments', which are repeated over and over again, in detail. And so let us be brief: As zero when multiplied by any finite number always gives zero, so the compilers could just as well have included 1000 such authors instead of 100 without the essence of their statements ever giving anything else but zero. They should accept that Relativity Theory cannot be condemned on the basis of an accumulation of 'judgements' by authors who have a certain command of the phraseology of Kant's critical philosophy but not the faintest idea of his spirit, just as the validity of Einstein's theories cannot be proved by majority resolutions of ladies' coffee parties.

A few sensible critics from the world of philosophy and physics have allowed themselves to be inserted between 'authors' of the latter type, and relativist scientists should not and in fact do not consider it beneath their dignity to cross swords with these. (Einstein himself, of course, as a pure researcher, is not fond of such scientific disputations!) But even here the fact that an opponent has a famous name unfortunately does not mean that he can be taken seriously. When one of these 'famous' men, for instance, is known to have only recently declared 'empirical astrology' (sic!) to be a science in the truest sense of the word, no physical scientist can be expected to embark upon a discussion about the justification of the conclusions of Relativity Theory with someone who does not even know the difference between science and dilettantism. And in fact it is he who writes sentences like:

In a quite impermissible manner, operations are carried out to prove that motion, which is assumed to be only relative, has an absolute real effect (shortening of scale, etc).

Pythian oracles, but not arguments against the crystal-clear logic of Einstein.

And so nothing remains of the criticism of the '100 authors' except a few significant objections by philosophers and physicists of keen judgement; Relativity Theory would naturally have to come to terms with these objections, had it not already successfully done so. And when for instance in carefully weighed sentences Professor Hartog, of Amsterdam, warns against extending 'relativization' to the workings of Nature as an *inner* experience or even to the field of ethical values, the creator of the Theory of Relativity would be the first to reject such an improper interpretation of his thought out of hand; even this most estimable contribution is fighting shadows. Taken as a whole the book is a product of such lamentable impotence that this regression into the 16th and 17th centuries can only be marvelled at and deplored. Only in politics does one meet such a depressing level; perhaps ideological antipathies are the only motive for this pamphlet.

Finally we must protest about the fact that in the bibliography authors are listed as being opponents of the Relativity Theory who have perhaps at some stage expressed certain misgivings about it, but who on the whole are definite adherents (Bottlinger, Poincaré, Prey)! One can only hope that German science is not shown up by such depressing rubbish again.

---

## Ultraviolet catastrophe

From *La Théorie du rayonnement et les quanta* (Solvay Conference) edited by Langevin and de Broglie (Paris 1912) p 77.

[*At the Solvay Conference of 1911, the subject of discussion was Radiation and Quanta. James Jeans tried to explain the ultraviolet catastrophe and the specific heat of solids in classical terms. He proposed a model in which each 'heat capacity' acted like a reservoir connected to others by a system of tubes and leaks. Poincaré's contribution to the discussion was brief.*]

It is obvious that by giving suitable dimensions to the communicating tubes between his reservoirs and giving suitable values to the leaks, Jeans can account for any experimental results whatever. But this is not the role of physical theories. They should not introduce as many arbitrary constants as there are phenomena to be explained; they should establish connections between different experimental facts, and above all they should allow predictions to be made.

# Flatland: a romance of many dimensions

From Nature
February 12,
1920.

From *Nature*
February 12,
1920.

[*An anonymous letter entitled 'Euclid, Newton, and Einstein,' published
in* Nature *on February 12, 1920, called attention to a little book by
Edwin Abbott Abbott (1838–1926), best known for his scholarly*
Shakespearian Grammar, *his life of Francis Bacon and a number of
theological discussions.*]

Some thirty or more years ago, a little *jeu d'esprit* was written by
Dr Edwin Abbott, entitled 'Flatland.' At the time of its publication
it did not attract as much attention as it deserved. Dr Abbott
pictures intelligent beings whose whole experience is confined to
a plane, or other space of two dimensions, who have no faculties
by which they can become conscious of anything outside that
space and no means of moving off the surface on which they live.
He then asks the reader, who has the consciousness of the third
dimension, to imagine a sphere descending upon the plane of
Flatland and passing through it. How will the inhabitants regard
this phenomenon? They will not see the approaching sphere and
will have no conception of its solidity. They will only be conscious
of the circle in which it cuts their plane. This circle, at first a
point, will gradually increase in diameter, driving the inhabitants
of Flatland outwards from its circumference, and this will go on
until half the sphere has passed through the plane, when the circle
will gradually contract to a point and then vanish, leaving the
Flatlanders in undisturbed possession of their country.

Their experience will be that of a circular obstacle gradually
expanding or growing, and then contracting, and they will attribute
to *growth in time* what the external observer in three dimensions
assigns to motion in the third dimension, through three-dimensional
space. Assume the past and future of the universe to be all depicted
in four-dimensional space and visible to any being who has consciousness of the fourth dimension. If there is motion of our three-
dimensional space relative to the fourth dimension, all the changes
we experience and assign to the flow of time will be due simply to
this movement, the whole of the future as well as the past always
existing in the fourth dimension.

From Edwin A
Abbott, *Flatland:
A Romance of
Many Dimensions*
(New York:
Barnes and Noble)
1963.

[*In a vision the narrator, a native of Flatland, has been indoctrinated by*
*Sphere to carry the Gospel of Three Dimensions to his blind benighted*
*countrymen in Flatland.*]

*I.* 'Pardon me, O Thou Whom I must no longer address as the
Perfection of all Beauty; but let me beg thee to vouchsafe thy
servant a sight of thine interior.'
*Sphere.* 'My what?'

*I.* 'Thine interior: thy stomach, thy intestines.'

*Sphere.* 'Whence this ill-timed impertinent request? . . .'

*I.* 'But my Lord has shewn me the intestines of all my countrymen in the Land of Two Dimensions by taking me with him into the Land of Three. What therefore more easy than now to take his servant on a second journey into the blessed region of the Fourth Dimension, where I shall look down with him once more upon this land of Three Dimensions, and see the inside of every three-dimensional house, the secrets of the solid earth, the treasures of the mines in Spaceland, and the intestines of every solid living creature, even of the noble and adorable Spheres'.

*Sphere.* 'But where is this land of Four Dimensions?'

*I.* 'I know not: but doubtless my Teacher knows'.

*Sphere.* 'Not I. There is no such land. The very idea of it is utterly inconceivable. . . . Men are divided in opinion as to the facts. And even granting the facts, they explain them in different ways. And in any case, however great may be the number of different explanations, no one has adopted or suggested the theory of a Fourth Dimension. Therefore, pray have done with this trifling, and let us return to business.'

---

## Schools of physics

From *Physicists continue to laugh* MIR Publishing House, Moscow 1968. Translated from the Russian by Mrs Lorraine T Kapitanoff.

When Niels Bohr visited the Physics Institute of the Academy of Sciences of the USSR, to the question of how he had succeeded in creating a first-rate school of physicists he replied: 'Presumably because I was never embarrassed to confess to my students that I am a fool . . .'.

On a later occasion, when E M Lifschitz read out this sentence from a translation of the speech it emerged in the following form: 'Presumably because I was never embarrassed to declare to my students that they are fools . . .'.

This sentence caused an animated reaction in the auditorium, then Lifschitz, looking at the text again, corrected himself and apologized for his accidental slip of the tongue. However, P L Kapitsa who had been sitting in the hall very thoughtfully noted that this was not an accidental slip of the tongue. It accurately expressed the principal difference between the schools of Bohr and of Landau to which E M Lifschitz belonged.

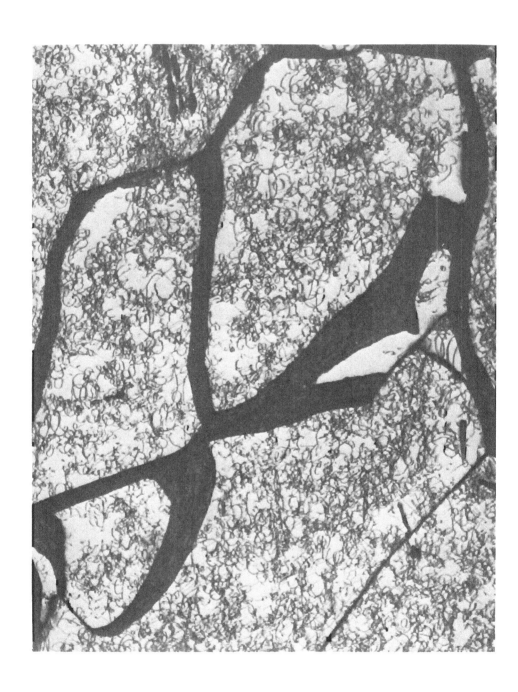

# How a theoretical physicist works

V BEREZINSKY

From *Paths into the Unknown* No 2. Printed in *Physicists continue to laugh* (Moscow: MIR Publishing House) 1968. Translated from the Russian by Mrs Lorraine T Kapitanoff.

I have always thought—although it was dangerous to express these thoughts aloud—that the theoretician has no role to play in physics. To say this in front of theoreticians is dangerous. They are convinced that experiments are needed only to verify the results of their theoretical calculations even though in reality everything is the other way around: laws are established experimentally and only then do theoreticians explain them.

But, as is well known, they can explain any result.

On one occasion we had completed an important experiment on the determination of the relationship between two physical quantities $A$ and $B$. I rushed to the telephone and called a famous theoretician who was occupied with the same problem.

'Volodya! We have finished. $A$ has turned out to be larger than $B$!'

'This is completely understandable. You didn't even have to make your experiment. $A$ is larger than $B$ for the following reasons. . . .'

'Oh no! Did I really say that $A$ was larger than $B$? I made a slip—it is $B$ which is larger than $A$!'

'Then this is even more understandable. Here is why. . . .' *

Unsuccessful experimenters usually become theoreticians. They notice even while they are students that if they simply remain for 5–10 minutes near any apparatus without even touching it, it is only fit for carrying straight to the dump. This follows them for their whole life. Once after a seminar, the famous German theoretician Sommerfeld said to his listeners 'And now let us take a look at how this apparatus, built on the principle we have worked out, operates.' The theoreticians trickled in single file after Sommerfeld into the laboratory, took off their spectacles and the knowledgable ones stared at the apparatus. Sommerfeld triumphantly turned on the switch . . . the apparatus burned up.

In the work of all theoreticians there is one common trait—they work differently. Don't get the idea that I want to say something good about their work; nothing could be further from my thoughts. Classical theoreticians worked with antiquated methods. They began work in flocks then dispersed into solitude along by-ways and paths and for hours, days, months gazed at everything that

---

* A story is told about Ia I Frenkel. It is said that in the Physical Theoretical Institute in the 30's, a certain experimenter caught up with him in a corridor and showed him a curve obtained in an experiment. After a minute's thought Ia I gave an explanation for the form of this curve. However it was explained that the curve had accidentally been turned upside down. The curve was put into place and having thought it over he explained this behaviour too.

met their eyes. A little sparrow chirped—they looked at the little sparrow; a fish splashed in the river, they lay down on their stomachs and watched the fish's path. Such a method was very pleasing to them because all theoreticians are terrible triflers but conceal it carefully. Call yourself a theoretician, and idleness becomes a strenuous contemplation of themes. You might think that this is not really so. You might believe for example that Newton was specially sitting under a tree and was awaiting the time when the apple would fall on him in order to discover the law of universal gravity. Not at all. He was simply shirking work. I am not saying that it was a bit ungentlemanly to discover the law thanks to an apple, but to claim the whole credit for himself was.

But in our times such a method of work is acknowledged as hopelessly antiquated. Now theoreticians prefer to begin working from the end. And this began with Einstein.

At the end of the 19th century the American physicist Michelson established experimentally (experimentally note), that it is impossible to catch a light ray. No matter how fast you ran after the ray it would always escape from you with a velocity of three hundred thousand kilometres per second.

Having rolled up his sleeves the theoretical physicist set to work: he placed an easy chair under the night sky and fixed an unblinking gaze on the shining stars. No matter how much he looked he was unable to give a sensible explanation of Michelson's experiment. But Einstein began from the end: he assumed that light possesses such a property and that was that. Theoreticians thought a little—some for ten years, others for twenty years, for as long as necessary, and then said: 'Brilliant!'

Whatever it was like previously, we see now that clear, direct and comprehensible experimental facts underlie theoretical work. Even in the midst of a piece of work, completely swamped and overshadowed with arguments and mathematical formulas, any theoretician can at once fish out of the sea of mathematics those conclusions which he intended to obtain from the very beginning. But it is by far the best if it is impossible to prove these conclusions experimentally.

In general, theoreticians adore examining effects which are unobservable in principle. For example, Dirac proposed that a uniform sea of electrons exists with negative energy which cannot be observed. But if one fishes out one electron from this sea, then a hole shows up in its place which we assume to be a positively charged electron—a positron.

Similar ideas were not new to Dirac as shown by one story which is still current at Cambridge. Dirac while still a student attended a mathematical congress where the following problem was set among others. The precise text of it is not available so I will paraphrase it.

Three fishermen were fishing on a secluded island. The fish briskly gobbled the bait; the fishermen were so absorbed that they did not notice that night had come and did not realise till too late what a mountain of fish they had hooked. So they had to spend the night on the island. Two fishermen quickly fell asleep, each nestled down under his boat, but the third had insomnia and decided to go home. He did not waken his comrades but divided all the fish into three parts. There proved to be one extra fish. After a moment's thought he threw it into the water, took his share and went home.

In the middle of the night the second fisherman woke up. He did not know that the first fisherman had already left and also divided all the fish into three and, as you might anticipate, there was one fish left over. This fisherman was not distinguished for his originality and he threw the fish in a little farther from the shore and with his share sat down in the boat. The third fisherman awoke toward morning. Not having washed and not having noticed that his comrades were no longer there he hastened to divide the fish. He divided them into three equal parts, threw the one extra fish into the water, took his share and that was that.

The problem was, to determine the least number of fish that the fishermen could have had.

Dirac proposed the following solution: there were $(-2)$ fishes. After the first fisherman carried out the antisocial action of throwing one fish into the water there were $(-2)-1 = -3$. Then he went, carrying in his bag $(-1)$ fish, and there were $(-3)-(-1) = -2$ fishes left behind. The second and third fishermen simply repeated the bad action of their comrade.

I could tell many other stories about theoreticians and their work, but they have told me that one theoretician is writing a story under the title 'How Experimental Physicists Work.' That, of course, will be presented upside down. He says that theoreticians predict all laws, and the experimenters merely confirm them. So I will hasten to conclude, only I don't know how to sign my name. I dare not use my real last name; how would I ever work afterwards? I could never discuss anything with a single theoretician. I will sign myself thus:

A well-meaning experimenter.

# The art of finding the right graph paper to get a straight line

S A RUDIN

Condensed from
*Journal of
Irreproducible
Results*, **12**
no 3 (1964).

As any fool can plainly see, a straight line is the shortest distance between two points. If, as is frequently the case, point A is where you are and point B is research money, it is most important to see to it that the line is as straight as possible. Besides, it looks more scientific. That is why graph paper was invented.

The first invention was simple graph paper, which popularized the straight line (figure 1). But people who had been working the constantly accelerating or decelerating paper had to switch to log paper (figure 2). If both coordinates were logarithmic, log–log paper was necessary (figure 3).

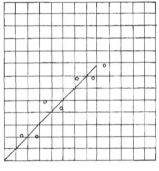

FIGURE 1

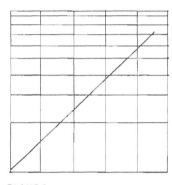

FIGURE 2

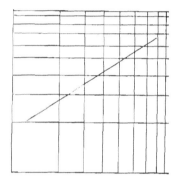

FIGURE 3

Or, if you had a really galloping variable on your hands, double log–log paper was the thing. And so on for all combinations and permutations of the above (figure 4).

FIGURE 4

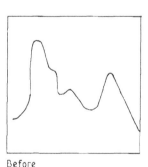

Before

After

For the statistician, there is always probability paper, which will turn a normal ogive into a straight line or a normal curve into

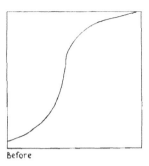
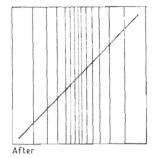

Before                    After

a tent. It is especially popular with statisticians, since it makes their work look precise (figure 5).

Sometimes correlation coefficient scattergrams come out at o·oo with a distribution shaped like a matzo ball (figure 6A). But using 'correlation paper' Pearson r's of any desirable degree of magnitude can be obtained (figure 6B). Naturally, negative correlation paper is available; it simply points the diagram the other way.

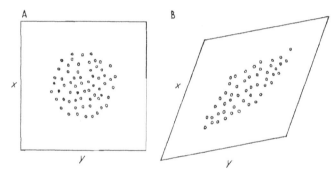

A                                    B

x                    x

y                         y

When you get a cycle where you should be getting a straight line, you use the following method. First, the peaks and troughs of the original plot are marked (figure 7A). Then, an overlay of transparent plastic sheet is put over it, and the dots alone copied. Now, it is obvious that these points are simply departures from a straight line, which is presented in dashed form (figure 7B). Finally, the straight line alone is recopied on to another graph paper (figure 7C).

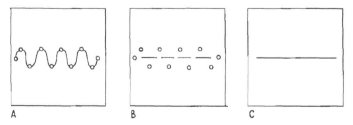

A                    B                    C

There is nothing so graphic as a graph to make a point graphically.

99

# On the imperturbability of elevator operators: LVII

S CANDLESTICKMAKER *Institute for Studied Advances, Old Cardigan, Wales* (Communicated by John Sykes; *received October 19, 1910*)

[*Professor John Sykes' famous spoof of Professor S Chandrasekhar delighted the 'victim', who arranged to have it printed in the format of* The Astrophysical Journal. *Some librarians bound it in series without noticing.*]

## ABSTRACT

In this paper the theory of elevator operators is completed to the extent that is needed in the elementary theory of Field's. It is shown that the matrix of an elevator operator cannot be inverted, no matter how rapid the elevation. An explicit solution is obtained for the case when the occupation number is zero.

## I. INTRODUCTION

In an earlier paper (Candlestickmaker 1954$q$; this paper will be referred to hereafter as 'XXXVIII') the simultaneous effect of a magnetic field, an electric field, a Marshall field, rotation, revolution, translation, and retranslation on the equanimity of an elevator operator has been considered. However, the discussion in that paper was limited to the case when incivility sets in as a stationary pattern of dejection; the alternative possibility of overcivility was not considered. This latter possibility is known to occur when a Marshall field alone is present; and its occurrence has been experimentally demonstrated by Shopwalker and Salesperson (1955) in complete disagreement with the theoretical predictions (Nostradamus 1555). The possibility of the occurrence of overcivility when no Marshall field is present has also been investigated (Candlestickmaker 1954$t$); and it has been shown that with substances such as U and I it cannot occur. It is therefore a matter of some importance that the manner of the onset of incivility be determined. This paper is devoted to this problem.

## 2. THE REDUCTION TO A TWELFTH-ORDER CHARACTERISTIC VALUE PROBLEM IN CASE OPERATORS A, B AND C ARE LOOKING IN THE SAME DIRECTION

The notation is more or less the same as in XXXVIII:

*Definitions*

$\gamma$ = first occupant,
$B_\eta$ = second occupant,
$g_g$ = third occupant,
$O$ = operator,
$\mathfrak{M}(O)$ = matrix of operator,

$a$ = acceleration of elevation of the conglomeration,
$\Omega_{2l}$ = critical Étage number for the onset of incivility,
$\Omega_{2l2} = \Omega_{2l}/\pi^{11/7}$.

The basic equations of the problem on hand are (cf. XXXVIII, eqs. [429] and [587])

$$\frac{\partial a}{\partial \beta} = \gamma \omega + n \nabla^2 j, \tag{1}$$

$$(5 + \pi)\, B_\eta = a + b + c, \tag{2}$$

$$x = x, \tag{3}$$

and

$$g_q + \tfrac{1}{2} m\, v^2 = 1. \tag{4}$$

Using also the relation (Pythagoras — 520)

$$3^2 + 4^2 = 5^2, \tag{5}$$

we find, after some lengthy calculations,

$$|\,\mathfrak{M}\,| = 0, \tag{6}$$

which shows that the matrix of the operator cannot be inverted. The required characteristic values $\Omega_{2l}$ are the solutions of equation (6). From the magnitude of the numerical work which was already needed for obtaining the solution for the purely rational case (cf. Candlestickmaker and Canna Helpit 1955) we may conclude that a direct solution of the characteristic value problem presented by equation (6) would be downright miraculous. Fortunately, as in XXXVIII, the problem can be solved explicitly in the case when the occupation number is zero. This is admittedly a case which has never occurred within living memory. However, from past experience with problems of this kind one may feel that any solution is better than none.

### 3. THE EQUATIONS DETERMINING THE MARGIN AT STATE IN THE CASE WHEN THE OCCUPATION NUMBER IS ZERO

For the reasons just given (i.e., because we cannot solve any other problem) we shall restrict ourselves in this paper to a consideration of the case when the occupation number is zero. In this case $\Omega_{2l}$ satisfies

$$\log \Omega_{2l} = 1, \tag{7}$$

the solution of which has been obtained numerically; it is approximately

$$\Omega_{2l} = 2\cdot 7. \tag{8}$$

This result shows that the transition to overcivility occurs between the values 2 and 3 given by Giftcourt (1956), respectively, Bookshelf (1956), a result which should be capable of direct experimental confirmation. The author hopes to deal with this problem next Saturday afternoon.

In conclusion, I wish to record my indebtedness to Miss Canna Helpit, who carried out the laborious numerical work involved in deriving equation (8).

The research reported in this paper has in part been suppressed by the Office of Navel Research under Contract Altum-OU812 with the Institute for Studied Advances.

REFERENCES

Bookshelf, M. F. 1956, *J. Gen. Psychol.*, **237**, 476.
Candlestickmaker, S. 1954a, *Zool. Jahrb.*, **237**, 476.
—. 1954b, *Parasitology*, **237**, 476.
——. 1954c, *Zentralbl. Bakt.*, **237**, 476.
——. 1954d, *Trans. N.-E. Cst Inst. Engrs. Shipb.*, **237**, 476.
——. 1954e, *R. C. Circ. mat. Palermo*, **237**, 476.
——. 1954f, *Adv. Sci.*, **237**, 476.
——. 1954g, *Math. Japonicae*, **237**, 476.
——. 1954h, *Biol. Bull. Woods Hole*, **237**, 476.
——. 1954i, *Bull. Earthq. Res. Inst. Tokyo*, **237**, 476.
——. 1954j, *J. Dairy Sci.*, **237**, 476.
——. 1954k, *Ann. Trop. Med. Parasitol.*, **237**, 476.
——. 1954l, *Trab. Lab. Invest. biol. Univ. Madrid*, **237**, 476.
——. 1954m, *Cellule*, **237**, 476.
——. 1954n, *Bot. Gaz.*, **237**, 476.
——. 1954o, *Derm. Zs.*, **237**. 476.
——. 1954p, *J. Pomol.*, **237**, 476.
——. 1954q, *Arch. Psychiat. Nervenkr.*, **237**, 476.
——. 1954r, *Sci. Progr. Twent. Cent.*, **237**, 476.
——. 1954s, *Portugaliae Math.*, **237**, 476.
——. 1954t, *Abh. senckenb. naturf. Gesellsch.*, **237**, 476.
Candlestickmaker, S., and Helpit, Canna E. 1955, *Compositio Math.*, **237**, 476.
Giftcourt, M. F. 1956, *J. Symbolic Logic*, **237**, 476.
Nostradamus, M. 1555, *Centuries* (Lyons).
Pythagoras—520, in: Euclid—300, *Elements*, Book I, Prop. 47 (Athens).
Shopwalker, M., and Salesperson, F. 1955, *Heredity*, **237**, 476.

Newspaper report North American Aviation has designed an atomic power plant . . . Designers of the model claim that burning 10 pounds of fissionable material in such a plant would produce as much power as the Hoover Dam.

# The analysis of contemporary music using harmonious oscillator wave functions

H J LIPKIN, *Department of Musical Physics, Weizmann Institute of Science*

Reprinted from *Proceedings of the Rehovoth Conference on Nuclear Structure,* held at the Weizmann Institute of Science, Rehovoth, September 8–14, 1957.

The importance of Harmonious Oscillation in music was well known [1] even before the discovery of the Harmonious Oscillator by Stalminsky [2]. Evidence for shell structure was first pointed out by Haydn [3], who discovered the magic number four and proved that systems containing four musicleons possessed unusual stability [4]. The concept of the magic number was extended by Mozart, who introduced the 'Magic Flute' [5], and a Magic Mountain was later introduced by Thomas Mann [6]. A system of four magic flutes is therefore doubly magic, and four magic flutes playing upon a magic mountain would be triply magic. Such a system is probably so stable that it does not interact with anything at all, and is therefore unobservable. This explains the fact that doubly and triply magic systems have never been observed.

A fundamental advance in the application of spectroscopic techniques to music is due to Racahmaninoff [7], who showed that all musical works can be expressed in terms of a small number of parameters, $A, B, C, D, E, F,$ and $G$, along with the introduction of Sharps [8]. Work along lines similar to that of Racahmaninoff has been done by Wigner, Wagner and Wigner [9] using the Niebelgruppentheorie. Relativistic effects have been calculated by Bach, Feshbach, and Offenbach, using the method of Einstein, Infeld, and Hoffman [10].

There has been no successful attempt thus far to apply the Harmonious Oscillator to modern music. The reason for this failure, namely that most modern music is not harmonious, was noted by Wigner, Wagner and Wigner [11].

A more unharmonious approach is that of Brueckner [12], who uses plane waves instead of harmonious oscillator functions. Although this method shows great promise, it is applicable strictly speaking only to infinite systems. The works of the Brueckner School are thus suitable only for very large ensembles.

A few very recent works should also be mentioned. There is the Nobel-Prize-winning work of Bloch [13] and Purcell [14] on unclear resonance and conduction. The work of Primakofiev should be noted [15], and of course the very fine waltzes presented by Strauss [16] at the 'Music for Peace' Conference in Geneva.

1 G F Handel, *The Harmonious Blacksmith* (London, 1757)

2 Igar Stalminsky, *Musical Spectroscopy with Harmonious Oscillator Wave Functions, Helv. Mus. Acta* 1 (1801) 1

3 J Haydn, *The α-particle of Music; the String Quartet Op 20* (1801) No 5

*The analysis of contemporary music using harmonious oscillator wave functions*

4 A B Budapest, C D Paganini and E F Hungarian, *Magic Systems in Music*

5 W A Mozart, *A Musical Joke*, K234567767 (1799)

6 T Mann, *Joseph Haydn and His Brothers* (Interscience, 1944)

7 G Racahmaninoff, *Sonority and Seniority in Music* (Invited Lecture, International Congress on Musical Structure, Rehovoth, 1957)

8 W T Sharp, *Tables of Coefficients* (Chalk River, 1955)

9 E Wigner, R Wagner and E P Wigner, *Der Ring Die Niebelgruppen. I Siegbahn Idyll* (Bayrut, 1900)

10 J S Bach, H Feshbach and J Offenbach, *Tales of Einstein, Infeld and Hoffman* (Princeton, 1944)

11 E P Wigner, R Wagner and E Wigner, *Gotterdammerung!!* and other unpublished remarks made after hearing 'Pierrot Lunaire'

12 A Brueckner, W Walton and Ludwig von Beethe, *Effective Mass in C Major*

13 E Bloch, *Schelomo, an Unclear Rhapsody*

14 H Purcell, *Variations on a Theme of Britten* (*A Young Person's Guide to the Nucleus*)

15 S Primakofiev, *Peter and the Wolfram-189*

16 J Strauss, *The Beautiful Blue Cerenkov Radiation*; *Scientist's Life*; *Wine, Women and Heavy Water*; *Tales from the Oak Ridge Woods*.

## Researchers' prayer

From *Proceedings of the Chemical Society*, January 1963, pp 8–10.

[*The Proceedings of the Chemical Society records some of the 128 verses submitted in a competition at Christmas 1962 for quatrains in the style of The Fisherman's Prayer: 'God give me strength to catch a fish/ So Large that even I/ When telling of it afterward/ May never need to lie.'*]

*Grant, oh God, Thy benedictions*
*On my theory's predictions*
*Lest the facts, when verified,*
*Show Thy servant to have lied.*

*May they make me B.Sc.,*
*A Ph. D. and then*
*A D. Sc., and F. R. S.,*
*A Times Obit. Amen.*

*Oh, Lord, I pray, forgive me please,*
*My unsuccessful syntheses,*
*Thou know'st, of course—in Thy position—*
*I'm up against such competition.*

*Let not the hardened Editor,*
*With referee to quote,*
*Cut all my explanation out*
*And print it as a Note.*

104

# Turboencabulator

J H QUICK

From The Institution of Electrical Engineers, Students' Quarterly Journal 25 (1955) p 184.

For a number of years now work has been proceeding in order to bring perfection to the crudely conceived idea of a machine that would not only supply inverse reactive current for use in unilateral phase detractors, but would also be capable of automatically synchronizing cardinal grammeters. Such a machine is the 'Turboencabulator.' Basically, the only new principle involved is that instead of power being generated by the relaxive motion of conductors and fluxes, it is produced by the modial interactions of magneto-reluctance and capacitive directance.

The original machine had a base-plate of prefabulated amulite, surmounted by a malleable logarithmic casing in such a way that the two spurving bearings were in direct line with the pentametric fan. The latter consisted simply of six hydrocoptic marzelvanes, so fitted to the ambifacient lunar vaneshaft that side fumbling was effectively prevented. The main winding was of the normal lotus-o-delta type placed in panendermic semiboloid slots in the stator, every seventh conductor being connected by a non-reversible termic pipe to the differential girdlespring on the 'up' end of the grammeter.

Forty-one manestically placed grouting brushes were arranged to feed into the rotor slip stream a mixture of high S-value phenyhydrobenzamine and 5 per cent reminative tetraiodohexamine. Both of these liquids have specific pericosities given by $p = 2 \cdot 5\ Cn$ where $n$ is the diathecial evolute of retrograde temperature phase disposition and $C$ is Cholmondeley's annual grillage coefficient. Initially, $n$ was measured with the aid of a metapolar pilfrometer, but up to the present date nothing has been found equal to the transcentental hopper dadoscope.

Electrical engineers will appreciate the difficulty of nubbing together a regurgitative purwell and a supraminative wennelsprocket. Indeed, this proved to be a stumbling block to further development until, in 1943, it was found that the use of anhydrous naggling pins enabled a kyptonastic bolling shim to be tankered.

The early attempts to construct a sufficiently robust spiral decommutator failed largely because of lack of appreciation of the large quasi-piestic stresses in the gremlin studs; the latter were specially designed to hold the roffit bars to the spamshaft. When, however, it was discovered that wending could be prevented by a simple addition of tooth to sockets almost perfect running was secured.

The operating point is maintained as near as possible to the HF rem peak by constantly fromaging the bituminous spandrels. This is a distinct advance on the standard nivelsheave in that no dram-mock oil is required after the phase detractors have remissed.

Undoubtedly, the turboencabulator has now reached a very high level of technical development. It has been successfully used for operating nofer trunnions. In addition, whenever a barescent skor motion is required, it may be employed in conjunction with a drawn reciprocating dingle arm to reduce sinusoidal deplenera-tion.

## Heaven is hotter than Hell

From *Applied Optics*, 11, A14 (1972).

The temperature of Heaven can be rather accurately computed from available data. Our authority is the Bible: Isaiah 30:26 reads, *Moreover the light of the Moon shall be as the light of the Sun and the light of the Sun shall be sevenfold, as the light of seven days.* Thus Heaven receives from the Moon as much radiation as we do from the Sun and in addition seven times seven (forty-nine) times as much as the Earth does from the Sun, or fifty times in all. The light we receive from the Moon is a ten-thousandth of the light we receive from the Sun, so we can ignore that. With these data we can compute the temperature of Heaven. The radiation falling on Heaven will heat it to the point where the heat lost by radiation is just equal to the heat received by radiation. In other words, Heaven loses fifty times as much heat as the Earth by radiation. Using the Stefan–Boltzmann fourth-power law for radiation

$$\left(\frac{H}{E}\right)^4 = 50,$$

where $E$ is the absolute temperature of the Earth—300K. This gives $H$ as 798 K (525°C).

The exact temperature of Hell cannot be computed but it must be less than 444·6°C, the temperature at which brimstone or sulphur changes from a liquid to a gas. Revelations 21:8: *But the fearful, and unbelieving . . . shall have their part in the lake which burneth with fire and brimstone.* A lake of molten brimstone means that its temperature must be below the boiling point, which is 444·6°C. (Above this point it would be a vapour, not a lake.)

We have, then, temperature of Heaven, 525°C. Temperature of Hell, less than 445°C. Therefore, Heaven is hotter than Hell.

# On the feasibility of coal-driven power stations

O R FRISCH

From *The Journal of Jocular Physics* 3, pp 27–30 in commemoration of the 70th birthday of Professor Niels Bohr (October 7, 1955) at the Institutet for Teoretick Fysick, Copenhagen.

The following article is reprinted from the Yearbook of the Royal Institute for the Utilization of Energy Sources for the Year MMMMCMLV, p1001.

In view of the acute crisis caused by the threat of exhaustion of uranium and thorium from the Earth and Moon Mining System, the Editors thought it advisable to give the new information contained in the article the widest possible distribution.

*Introduction.* The recent discovery of coal (black fossilized plant remains) in a number of places offers an interesting alternative to the production of power from fission. Some of the places where coal has been found show indeed signs of previous exploitation by prehistoric men who, however, probably used it for jewels and to blacken their faces at tribal ceremonies.

The power potentialities depend on the fact that coal can be readily oxidized, with the production of a high temperature and an energy of about 0·0000001 megawattday per gramme. This is, of course, very little, but large amounts of coal (perhaps millions of tons) appear to be available.

The chief advantage is that the critical amount is very much smaller for coal than for any fissile material. Fission plants become, as is well known, uneconomical below 50 megawatts, and a coal-driven plant may be competitive for isolated communities with small power requirements.

*Design of a coal reactor.* The main problem is to achieve free, yet controlled, access of oxygen to the fuel elements. The kinetics of the coal-oxygen reaction are much more complicated than fission kinetics, and not yet completely understood. A differential equation which approximates the behaviour of the reaction has been set up, but its solution is possible only in the simplest cases.

It is therefore proposed to make the reaction vessel in the form of a cylinder, with perforated walls to allow the combustion gases to escape. A concentric inner cylinder, also perforated, serves to introduce the oxygen, while the fuel elements are placed between the two cylinders. The necessary presence of end plates poses a difficult but not insoluble mathematical problem.

*Fuel elements.* It is likely that these will be easier to manufacture than in the case of fission reactors. Canning is unnecessary and indeed undesirable since it would make it impossible for the oxygen to gain access to the fuel. Various lattices have been calculated, and it appears that the simplest of all—a close packing of

107

equal spheres—is likely to be satisfactory. Computations are in progress to determine the optimum size of the spheres and the required tolerances. Coal is soft and easy to machine; so the manufacture of the spheres should present no major problem.

*Oxidant.* Pure oxygen is of course ideal but costly; it is therefore proposed to use air in the first place. However it must be remembered that air contains 78 per cent of nitrogen. If even a fraction of that combined with the carbon of the coal to form the highly toxic gas cyanogens this would constitute a grave health hazard (see below).

*Operation and Control.* To start the reaction one requires a fairly high temperature of about 988°F; this is most conveniently achieved by passing an electric current between the inner and outer cylinder (the end plates being made of insulating ceramic). A current of several thousand amps is needed, at some 30 volts, and the required large storage battery will add substantially to the cost of the installation.

There is the possibility of starting the reaction by some auxiliary self-starting reaction, such as that between phosphine and hydrogen peroxide; this is being looked into.

Once the reaction is started its rate can be controlled by adjusting the rate at which oxygen is admitted; this is almost as simple as the use of control rods in a conventional fission reactor.

*Corrosion.* The walls of the reactor must withstand a temperature of well over a 1000°F in the presence of oxygen, nitrogen, carbon monoxide and dioxide, as well as small amounts of sulphur dioxide and other impurities, some still unknown. Few metals or ceramics can resist such gruelling conditions. Niobium with a thin lining of nickel might be an attractive possibility, but probably solid nickel will have to be used. For the ceramic, fused thoria appears to be the best bet.

*Health Hazards.* The main health hazard is attached to the gaseous waste products. They contain not only carbon monoxide and sulphur dioxide (both highly toxic) but also a number of carcinogenic compounds such as phenanthrene and others. To discharge those into the air is impossible; it would cause the tolerance level to be exceeded for several miles around the reactor.

It is therefore necessary to collect the gaseous waste in suitable containers, pending chemical detoxification. Alternatively the waste might be mixed with hydrogen and filled into large balloons which are subsequently released.

The solid waste products will have to be removed at frequent intervals (perhaps as often as daily!), but the health hazards involved in that operation can easily be minimized by the use of conventional remote-handling equipment. The waste could then be taken out to sea and dumped.

There is a possibility—though it may seem remote—that the oxygen supply may get out of control; this would lead to melting of the entire reactor and the liberation of vast amounts of toxic gases. Here is a grave argument against the use of coal and in favour of fission reactors which have proved their complete safety over a period of several thousand years. It will probably take decades before a control system of sufficient reliability can be evolved to allay the fears of those to whom the safety of our people is entrusted.

## Bedside manner

From a first year examination in Physics for medical students
Q: Explain in molecular terms why hot air rises.
A: When a gas is heated, the molecules move faster. By Einstein's theory of relativity, the mass of a body increases with velocity.

$$\text{Density} = \text{mass/volume}$$

The mass increases $\therefore$ the density decreases, so the hot air rises.

Sir Arthur Eddington, quoted in *Astrophysical Journal* 101, 133 (1945). When an investigator has developed a formula which gives a complete representation of the phenomena within a certain range, he may be prone to satisfaction. Would it not be wiser if he should say 'Foiled again! I can find out no more about Nature along this line.'

'The fusion plasma requires a temperature of 500 million degrees, but I forget whether that's Centigrade or Absolute.'—Remark overheard by Arthur H Snell, Oak Ridge National Laboratory

# A theory of ghosts

D A WRIGHT

From *The Worm Runner's Digest* 12, 95 (1971).

It should be stated at the outset that this is a paper on physics, not metaphysics. Many physicists have turned to pseudo-philosophy, metaphysics or parapsychology, but not the writer; at least not yet.

It is well known that ghosts can penetrate closed doors and internal walls of buildings up to four inches or so (0·1 m) in thickness. There is some evidence however that they remain confined when present in old buildings with external wall thickness of a foot or more. According to the elementary ideas of wave mechanics (Schrödinger 1928, de Broglie and Brillouin 1928) this establishes them as objects whose associated wave functions decrease to $1/2 \cdot 7$ of their full amplitude at about 0·1 m from their boundary. Their wavelength is therefore of this order of magnitude and their mass at low velocity must be less than that of the electron by a factor of the order of $10^{16}$, that is it must be about $10^{-46}$ kg.

Evidently an object of such low mass can be accelerated to high velocity with very little expenditure of energy. Relativistic effects must therefore be considered when dealing with its motion (Einstein 1905) and it will be understood that velocities such as the escape velocity from the earth's gravitational field can readily be attained. The latter velocity is 25 000 mph, or 10 km s$^{-1}$, independent of the mass of the object (Newton 1687). The energy required is only $10^{-38}$ J. A breath of wind will therefore more than suffice to start the ghost on a journey through the solar system, while minor interactions en route could eject it from the solar system on the way to the stars. The recently discovered solar wind (Cowley 1969) will suffice to accelerate ghosts almost to the velocity of light away from the sun's neighbourhood.

It is not surprising that in spite of the enormous number of ghosts formed by the demise of *homo sapiens* alone, over the last million years or so, the number of ghosts encountered on the earth's surface remains small. Admittedly, it is not obvious that *homo sapiens* is the only source of ghostly objects. It is likely, however, that all ghost material has extremely low density, so that the ghosts of large objects, both animate and inanimate, will also be dispersed very readily [1]. However, to pursue this topic

1 A collision between two cars recently reported in the press caused the one to disappear and no damage to the other. Clearly a ghost car of very low mass was involved.

would be an unwarrantable digression from the main subject of interest to us, which is naturally the ghost of human origin [2].

HUMAN GHOSTS

Proceeding with this subject, it is clear that when, for example, a person is pierced with a spear which is not removed, or hanged in chains, his ghost will remain at the spot and haunt it even though the sad event occurs in the open air. The spear or the chains are real objects with normal mass. In the absence of such impedimenta, a ghost will however rapidly leave the site and as we have seen will probably leave the earth and will quite possibly leave the solar system. However, following death in dungeons or in the interior of old castles with thick walls and small windows, the escape probability is very small even with the small mass we have determined and the ghost will haunt such a habitat for many years. (Wearing armour or dragging chains will of course prolong the period enormously. A layer of dust will produce a substantial increase.)

It is interesting to notice that a ghost will be accelerated to, say, 0·7 times the velocity of light by a very small amount of work, about $10^{-29}$ J. Its mass is then twice its rest mass (see Einstein 1905) and its wavelength is halved (see Schrödinger 1928, de Broglie and Brillouin 1928). Thus it is *less* able to penetrate a wall or door once its speed has increased substantially. A ghost in rapid motion in a confined space therefore will be less likely to escape than when moving slowly. It will also be difficult to locate. Although its momentum will be small, it will be large enough to displace light-weight objects on collision. Thus, we have an explanation not only of ghosts themselves, but also of the 'poltergeist' phenomenon; vases and other light articles will be displaced from shelves in a, no doubt, disconcerting manner, since the presence of high speed ghosts will be almost impossible to observe directly. Like many so called 'elementary particles' in physics, their presence can be detected only by the secondary effects they produce (for example, the neutrino, see Pauli 1933).

Evidently one can in no sense eject such ghosts by the use of violence; any further increase in an already high velocity will merely make escape more difficult. The only approach, if the

2 Consider, for example, A Pope, *An Essay on Mankind*.

presence of a high speed ghost is deemed undesirable, is to seek to calm it and bring it to rest, so that it can glide slowly through the wall. No doubt the procedure of exorcism is intended to achieve this result, though the author confesses that the details of how this is done remain obscure to him. It follows, incidentally, as will be seen below that the attempt is best made in near darkness.

It should be realized that the velocity of ghosts due to thermal agitation will be very large at ordinary temperatures, in view of their remarkably small mass. Thus the average energy of 20°C, $3kT/2$ (Maxwell 1860, Boltzmann 1872), will correspond with a velocity near that of light. Few ghosts will be moving slowly enough to be seen, unless they are very cold, or they attach themselves to some material object.

OBSERVATION OF GHOSTS

When light impinges on the surface of an object, it exerts pressure (Maxwell 1873) and carries momentum. One photon of visible light incident on the surface of a ghost and reflected from it could transfer momentum $2h\nu/c$, $10^{-27}$ J s m$^{-1}$, which would cause acceleration to a very high velocity. A ghost which was not loaded, or holding on to some object or person, would be removed rapidly if the walls were thin, or would otherwise display poltergeist phenomena. Presumably the reflection coefficient of the surface of a ghost must be much less than 100%, or it might never be seen at all. No doubt for this reason it appears to be general experience that ghosts are seen only under conditions of poor illumination. To examine a ghost, one should not shine a torch at it; a shielded candle is more suitable.

The low mass leads to a very large shift in wavelength $\Delta\lambda$ of radiation incident on a ghost's surface and scattered by it (Compton 1923). The value of $\Delta\lambda$ for a mass of $10^{-46}$ kg can be as large as $10^4$ m; thus all short wave radiation such as light, infrared *etc* will be scattered at radiofrequencies. The scattering of short wave radiation by ghosts in flight through the universe will therefore be a major source of cosmic radio noise. Attempts made so far by astronomers to explain this noise have, unfortunately, taken too little account of this contribution.

It has sometimes been thought that ghosts produce a sensation of cold in their environment. This is perhaps to be expected if they have just returned from outer space, where the temperature is believed to be about three degrees absolute (Penzias and Wilson 1965). It is less obvious why this should occur if they have been resident for some time, as in an old castle (unless, indeed, they have internal means of refrigeration, which seems unlikely, but perhaps not impossible). If the observation is correct, it implies that ghosts must have quite a high specific heat. This would, in turn, indicate that in spite of their very low mass, they are not structureless objects. It is evidently important to obtain more reliable evidence as to the temperature and specific heat of ghosts [3].

STATISTICS OF GHOSTS

The concept of a quasiparticle of large area and volume is new to physics, though it does not appear to be excluded *a priori*. Whether such an object would seem to us to be hot or cold when stationary is not by any means obvious; temperature and specific heat of ordinary particles depend on the state of motion. Thus, even if the observation is correct, it is not certain that ghosts have structure; they might still be elementary particles. Moreover the observation may be wrong; the impression of cold may be an illusion, or the result of faulty reporting. It is conceivable, for instance, that the observer experiences a sensation of cold through fear, although it is not obvious why such a reaction occurs.

We shall, therefore, proceed to consider the situation if the concept of a quasiparticle is applicable. It would then be desirable to investigate their spin properties, which would determine whether they obey the Fermi–Dirac (Fermi 1926, Dirac 1926) or the Bose–Einstein (Bose 1924, Einstein 1924) statistics [4]. It is always desirable in physics (both pure and applied), when a new question emerges, to propose an experimental method of obtaining evidence. In this case the behaviour of ghosts in a magnetic field would be enlightening [5].

If ghosts tend to accumulate in any part of the universe, in what the physicist would no doubt call a 'sink', and if they can be regarded as particles, they will constitute a 'degenerate' or 'condensed' population even at very low density (eg one ghost per metre$^3$. The details will of course depend on which statistics they obey.

3 Their measurement might constitute a valuable practical project for final year university students in applied physics.
4 All readers will appreciate the importance of this issue.
5 This might also lead to a good final year project. One is always on the lookout for these.

It is tempting to envisage that in human ghosts (and indeed not only human) a trace of sexual difference is 'carried over' [6], and would be represented by the antisymmetric wavefunctions characteristic of Fermi–Dirac statistics. Particles obeying these statistics would have half integral spin, and the ultimate state would be one in which ghosts of opposite spin had paired up to occupy the energy states available, each pair in one state. This highly satisfactory disposition from the point of view of the physicist [7] might well constitute a state of bliss that all ghosts hope to achieve.

Whether there is such a 'pool' or 'sink', whether these terms are really appropriate for such a state, and where in the universe it is, remain problems which we may solve only in the future. Meanwhile, we have offered a theory which meets the requirements of a contribution to science: It coordinates the known facts in the light of existing knowledge, it is not contrary to known facts and it suggests further lines of enquiry to be pursued in the future. Furthermore, it illuminates an area of human experience that had previously been thought inaccessible to the scientific method.

6 Here I confess we enter the realm of speculation, not really consistent with a scientific discourse.

7 The mathematician no doubt would refer to it as an 'elegant' solution.

FURTHER READING

Boltzmann L 1872 *Wien. Ber.* **66** 275
Bose S N 1924 *Z. Phys.* **26** 178
Compton A H 1923 *Phys. Rev.* **22** 411
Cowley T G 1969 *The Observatory* **89** 217
De Broglie L and Brillouin L 1928 *Selected Papers on Wave Mechanics* (London: Blackie)
Dirac P A M 1926 *Proc. R. Soc.* A **112**
Einstein A 1905 *Ann. Phys. Lpz.* **17** 891
Einstein A 1924 *Sitzber. Preuss. Akad. Wiss. Berlin* **261**
Fermi E 1926 *Z. Phys.* **36** 902
Maxwell J C 1860 *Phil. Mag.* **19** 19–21
Maxwell J C 1873 *A Treatise on Electricity and Magnetism* (Oxford: Clarendon Press)
Newton I 1687 *Principia* London
Pauli W 1933 *Inst. Solvay, 7th Session, Brussels* p324
Penzias A A and Wilson R W 1965 *J. Astrophys.* **142** 419
Schrödinger E 1928 *Collected Papers on Wave Mechanics* (London: Blackie)

# A stress analysis of a strapless evening gown

ANON

Condensed from
*A Stress Analysis
of a Strapless
Evening Gown
and other essays,*
ed Robert A Baker
(Prentice-Hall)
1963.

Effective as the strapless evening gown is in attracting attention, it presents tremendous engineering problems to the structural engineer. He is faced with the problem of designing a dress which appears as if it will fall at any moment and yet actually stays up with some small factor of safety. Some of the problems faced by the engineer readily appear from the following structural analysis of strapless evening gowns.

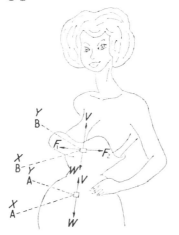

FIGURE 1. Forces acting on cloth element.

If a small elemental strip of cloth from a strapless evening gown is isolated as a free body in the area of plane A in figure 1, it can be seen that the tangential force $F$ is balanced by the equal and opposite tangential force $F$. The downward vertical force $W$ (weight of the dress) is balanced by the force $V$ acting vertically upward due to the stress in the cloth above plane A. Since the algebraic summation of vertical and horizontal forces is zero and no moments are acting, the elemental strip is at equilibrium.

Consider now an elemental strip of cloth isolated as a free body in the area of plane B of figure 1. The two tangible forces $F_1$ and $F_2$ are equal and opposite as before, but the force $W$ (weight of dress) is not balanced by an upward force $V$ because there is no cloth above plane $B$ to supply this force. Thus, the algebraic summation of horizontal forces is zero, but the sum of the vertical forces is not zero. Therefore, this elemental strip is not in equilibrium; but it is imperative, for social reasons, that this elemental strip be in equilibrium. If the female is naturally blessed with sufficient pectoral development, she can supply this very vital

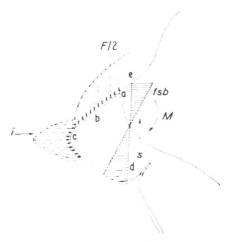

FIGURE 2. Force distribution of cantilever beam
(fsb = flexural stress in beam.)

force and maintain the elemental strip at equilibrium. If she is not, the engineer has to supply this force by artificial methods.

In some instances, the engineer has made use of friction to supply this force. The friction force is expressed by $F = fN$, where $F$ is the frictional force, $f$ the coefficient of friction and $N$ is the normal force acting perpendicularly to $F$. Since, for a given female and a given dress, $f$ is constant, then to increase $F$, the normal force $N$ has to be increased. One obvious method of increasing the normal force is to make the diameter of the dress at c in figure 2 smaller than the diameter of the female at this point. This has, however, the disadvantage of causing the fibres along the line c to collapse, and, if too much force is applied, the wearer will experience discomfort.

As if the problem were not complex enough, some females require that the back of the gown be lowered to increase the exposure and correspondingly attract more attention. In this case, the horizontal forces $F_1$ and $F_2$ (figure 1) are no longer acting horizontally, but are replaced by forces $T_1$ and $T_2$ acting downward at an angle $a$. Therefore, there is a total downward force equal to the weight of the dress below $B +$ the vector summation of $T_1$ and $T_2$. This vector sum increases in magnitude as the back is lowered because $R = 2T \sin a$, and the angle $a$ increases as the back is lowered. Therefore, the vertical uplifting force which has to be supplied for equilibrium is increased for low-back gowns.

Since these evening gowns are worn to dances, an occasional horizontal force, shown in figure 2 as $i$, is accidentally delivered to the beam at the point c, causing impact loading, which compresses all the fibres of the beam. This compression tends to cancel the tension in the fibres between e and b, but it increases the compression between c and d. The critical area is at point d, as the fibres here are subject not only to compression due to moment and impact, but also to shear due to the force $S$; a combination of low, heavy dress with impact loading may bring the fibres at point d to the 'danger point.'

There are several reasons why the properties discussed in this paper have never been determined. For one, there is a scarcity of these beams for experimental investigation. Many females have been asked to volunteer for experiments along these lines in the interest of science, but unfortunately, no cooperation was encountered. There is also the difficulty of the investigator having the strength of mind to ascertain purely the scientific facts. Meanwhile, trial and error and shrewd guesses will have to be used by the engineer in the design of strapless evening gowns until thorough investigations can be made.

## Two classroom stories

ROBERT WEINSTOCK

One day when using an audio oscillator in a lecture demonstration, I asked students to raise their hands if they could hear the note being emitted from the loudspeaker. As I increased the frequency, the hands began to drop; eventually all of them dropped as the pitch passed the human threshold. 'But isn't there anyone who can hear the note?' I persisted. From one of the students at the rear came the reply, 'Bow-wow!'

Paul Kirkpatrick was giving a lecture demonstration on X-rays. He asked for a handbag for interposition between the X-ray source and a fluorescent screen; one co-ed lent hers. In the darkened room he substituted another purse with a pistol inside. The co-ed let out a shriek of embarrassment, the rest of the class a roar of laughter. (But that was during the domestically tranquil days of World War II; perhaps the stunt would not be funny today.)

# Murphy's law

D L KLIPSTEIN

Condensed from 'The Contribution of Edsel Murphy to the Understanding of Behaviour in Inanimate Objects' in EEE: The Magazine of Circuit Design (August 1967).

Murphy's Law states that 'If anything can go wrong, it will.' Or, to state it in more exact mathematical form:

$$1 + 1 \; \text{☞} \; 2$$

where ☞ is the mathematical symbol for 'hardly ever'.

To show the all-pervasive nature of Murphy's work, the author offers a few applications of the law to the electronic engineering industry.

## GENERAL ENGINEERING

1. A patent application will be preceded by one week by a similar application made by an independent worker.

2. The more innocuous a design change appears, the further its influence will extend.

3. All warranty and guarantee clauses become void on payment of invoice.

4. An important Instruction Manual or Operating Manual will have been discarded by the Receiving Department.

## MATHEMATICS

5. Any error that can creep in, will. It will be in the direction that will do most damage to the calculation.

6. All constants are variables.

7. In a complicated calculation, one factor from the numerator will always move into the denominator.

## PROTOTYPING AND PRODUCTION

8. Any wire cut to length will be too short.

9. Tolerances will accumulate unidirectionally towards maximum difficulty of assembly.

10. Identical units tested under identical conditions will not be identical in the field.

11. If a project requires $n$ components, there will be $(n-1)$ units in stock.

12. A dropped tool will land where it can do most damage; the most delicate component will be the one to drop. (Also known as the principle of selective gravity.)

13. A device selected at random from a group having 99 per cent reliability will be a member of the 1 per cent group.

14. A transistor protected by a fast-acting fuse will protect the fuse by blowing first.

15. A purchased component or instrument will meet its specifications long enough, and only long enough, to pass Incoming Inspection.

16. After an access cover has been secured by 16 hold-down screws, it will be discovered that the gasket has been omitted.

## Thermoelectric effect

[*We should never laugh at the scientific speculations of past ages, but we can surely be permitted a smile. These extracts are from the notebook (now in the possession of L Mackinnon, Essex University) of a student who attended lectures on Natural Science at Marischal College, Aberdeen (one of the two universities then in that town) in the academic year 1834–5. Thermoelectricity had been discovered twelve years previously.*]

The electrical motion produced by heating a body, is produced by the Sun on the Earth. The effect of heat on matter can produce Electricity in different ways. The course of the Sun appears to be from East to West, and this is occasioned by the evolution of the Magnetic action. We already mentioned that Mr Sebeck heated one of the slips of copper, hence as it is warmer at the Equator than at the North pole, we may consider the Equator the heated part, and the North pole that which is kept cool. Electricity is the cause of bodies revolving round the Sun, and by this means we are enabled to explain the causes of the revolutions of all the heavenly bodies.

With regard to volcanos. The Earth is composed of Strata and admitting the supposition that the Earth is a large Galvanic battery, we shall be able to account for several phenomenas in nature.

The production of fire in the bowels of the Earth is evolved from the Galvanic apparatus, without being supplied from any other quarter.

There is a burning mountain in Iceland, and when Veṣuvius or Ætna are in eruption, they are in all probability the termination of a large pile. The compass needle is considerably affected by a 'Thunder Storm' for the Electricity of the atmosphere carries off the Magnetism, and by the 'Eruption of volcanos'.

# A glossary for research reports

C D GRAHAM, JR.

From *Metal Progress* 71, 75 (1957).

| | |
|---|---|
| It has long been known that. . . . | I haven't bothered to look up the original reference |
| . . . of great theoretical and practical importance | . . . interesting to me |
| While it has not been possible to provide definite answers to these questions . . . | The experiments didn't work out, but I figured I could at least get a publication out of it |
| The W–Pb system was chosen as especially suitable to show the predicted behaviour. . . . | The fellow in the next lab had some already made up |
| High-purity . . . <br> Very high purity . . . <br> Extremely high purity . . . <br> Super-purity . . . <br> Spectroscopically pure . . . | Composition unknown except for the exaggerated claims of the supplier |
| A fiducial reference line . . . | A scratch |
| Three of the samples were chosen for detailed study . . . | The results on the others didn't make sense and were ignored |
| . . . accidentally strained during mounting | . . . dropped on the floor |
| . . . handled with extreme care throughout the experiments | . . . not dropped on the floor |
| Typical results are shown . . . | The best results are shown |
| Although some detail has been lost in reproduction, it is clear from the original micrograph that . . . | It is impossible to tell from the micrograph |
| Presumably at longer times . . . | I didn't take time to find out |
| The agreement with the predicted curve is excellent | fair |
| good | poor |
| satisfactory | doubtful |
| fair | imaginary |

| | |
|---|---|
| .. as good as could be expected | non-existent |
| These results will be reported at a later date | I might possibly get around to this sometime |
| The most reliable values are those of Jones | He was a student of mine |
| It is suggested that ... It is believed that ... It may be that ... | I think |
| It is generally believed that ... | A couple of other guys think so too |
| It might be argued that ... | I have such a good answer to this objection that I shall now raise it |
| It is clear that much additional work will be required before a complete understanding ... | I don't understand it |
| Unfortunately, a quantitative theory to account for these effects has not been formulated | Neither does anybody else |
| Correct within an order of magnitude | Wrong |
| It is to be hoped that this work will stimulate further work in the field | This paper isn't very good, but neither are any of the others in this miserable subject |
| Thanks are due to Joe Glotz for assistance with the experiments and to John Doe for valuable discussions | Glotz did the work and Doe explained what it meant |

---

Technological progress has merely provided us with more efficient means for going backwards.

ALDOUS HUXLEY

# Why we must go to the Moon

CHARLES G TIERNEY

From Martin Levin's *Phoenix Nest*, *Saturday Review* (13 December, 1969), p 4.

Many persons have asked me, why should we send men to the moon. These concerned citizens question the wisdom of spending billions to explore space, when so much remains to be done here on earth in combating acid indigestion and dull, unmanageable hair. To these people I give a simple answer: We need the data that the moon and the planets can provide. And we need it pretty quick.

It is no longer a secret that the world's resources of unprocessed data are running dangerously low. Experts estimate variously that known reserves, once so abundant, will be exhausted in five to fifteen years. Unless new supplies are found before then, a crisis of unprecedented proportions will be upon us. To a world running out of raw facts, the moon promises a vast, untapped mine of new information, never before punched on cards, and sufficient to take the pressure off the situation for decades. SURVEYOR II revealed what appear to be natural lumps of pure data the size of turtle eggs lying exposed on the surface, waiting to be scooped up. So rich a trove so near at hand makes the moon our best hope for staving off a dilemma that becomes yearly more acute. Indeed, the spectacle of a grown man travelling 250000 miles to gather a sackful of pebbles takes on meaning only when we consider it in this light.

Still, 'I don't get it', some troubled questioners persist, and their artless query strikes close to the heart of the issue. For few laymen are able to appreciate the danger toward which we are swiftly drifting. In a few short years, all the data the earth has to offer will have been ground through the world-wide array of data-processing machines; all the computations possible will have been performed, analysed, printed out, and stored. Eventually, one by one, the tape reels will come to a halt, the control units will cease their clicking, the flickering console lights will fall into a steady, ominous pattern. Computer centres everywhere will be suffused with the dull reddish glow of a thousand warning FEED lights demanding input.

Unless we go to the moon now, there will be no input to feed them. The thought has given men at the Rand Corporation the cold shivers.

Why is this so? That is a question the experts seem reluctant to talk about. 'A busy computer is a contented computer,' they murmur. And indeed, extravagant measures are taken to protect the big brains against the possibility of idleness. They are kept running around the clock, watched by relays of operators oriented to scramble for fresh material when the FEED light glows.

Originally, this began as a matter of economic utilization of costly equipment. But it has long since gone far beyond that simple concept. Again, the experts are vague. 'The devil finds work for idle circuits to do,' they are apt to mutter, uneasily. This topic meets everywhere with ill-concealed anxiety and evasiveness, and a chilling conclusion eventually forces itself upon one: at bottom, nobody knows—nobody really knows for sure how the computers would react if the data stopped coming.

'They ask for data—we give it to them,' snapped a dean at MIT. 'We don't want no trouble.'

'They are very smart cookies,' said an IBM engineer carefully. 'Their memories are exhaustive, their logic is infallible, their decisions are—ruthless.' He hesitated. 'They do not know compassion.' Then he clammed up.

The importance and urgency of gaining access to the lunar data fields is apparent in the vast amount of money, effort, and risk involved in bringing it about. The conclusion is inescapable that, not only does no one know what to expect from a population of computers contemplating starvation—no one cares to find out. The expedition to the moon is a gigantic undertaking, fraught with peril and demanding of much sacrifice. But there is little choice. We must go.

---

### Face to face with metrication

NORMAN STONE *Chief Information Officer, Metrication Board, London*

Some traditional weights and measures are funny enough in themselves. I cannot say 'Two fardels equal one nooke' to myself without smiling; I am delighted that the fathom originally meant the distance a Viking encompassed in a hug; the Statute of Henry I which defined the foot in terms of thirty-six barleycorns taken from the width of the ear has a charm of its own; there is something laughable in the fact that the gauge of railways in Britain is the same as the distance between the wheels of a Roman chariot; and who would not be amused at the recollection that the basis of much modern town planning is the acre, an area ploughable in one day by a team of two oxen. It is one furrow long (furlong) by one chain.

# Life on Earth (by a Martian)

PAUL A WEISS

Condensed from *The Rockefeller Institute Review* 2 no 6, pp 8–14. Illustrations by Mrs Vera Teleki.

We had timed our approach to Earth to fall into the dark phase of the diurnal cycle. Yet, as we came nearer, we noticed that not all was darkness. There were streaks of light like knotted ribbons. They seemed to move in waves, mostly in one direction. From a still closer view, the knots proved to be separate bodies; they

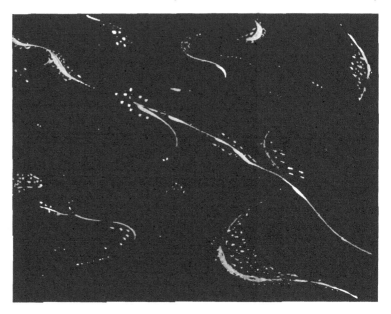

moved indeed, in spasms of alternating spurts and stalls. If life is motion, here was life. Each luminous knot was obviously an individual organism. Each had the polarized appearance of a wedge and seemed in finer resolution to consist of smaller subunits; something like cells. Why they should move in file and unison did not become clear to us until we came in still closer and saw the cause: they all were positively phototactic, attracted by a flickering light source, toward which we likewise now steered our course. Hovering over it, we recognized our first mistake: the knots, which from a distance had looked to us like single individuals, had dissociated into the putative 'subunits.' These latter thus revealed themselves, by their separate existence and independent motility, as the true elemental carriers of life on Earth.

So we kept watching them. Rather than rush right on into the attractive light source, they stopped just short of it and formed a handsome sort of crystalline array. For some time after, they remained immobile and soundless. The only sounds we heard

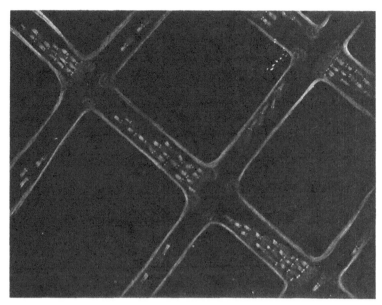

came from that light trap. We thought they slept. So, we took
courage to land and inspect them. That was when we noticed that
motion had stopped only superficially. For inside each unit we
perceived a pair of structures that kept on squirming. Whether
these were peristaltic internal organs or some sort of wiggly

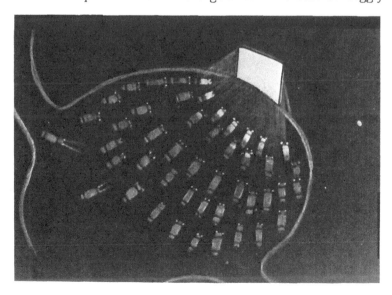

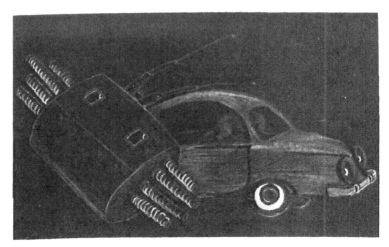

endoparasites, we had to leave unresolved, for our inspection was cut short abruptly when suddenly the light went out and the various creatures resumed locomotion, all in the reverse direction, though not all at once.

In order to study life on Earth in more detail, we decided to wait for daylight, fully prepared to be discovered. Unaccountably, this never happened. We therefore were able to go about our explorations undisturbed and at great leisure. And what we

learned was truly unexpected, almost incredible. To sum it up: Life on Earth has so many features in common with our own that we must concede the possibility that Martians and 'Earthians,' as we call them, might be akin. Unpalatable though this deduction may be to the glorious leaders of our superior Martian people, in our conscience as scientists we cannot evade it, provided we can document its premise. And document it we can. For we can prove that Earthians metabolize, eat, drink, and groom; generate heat and light and sound; proliferate and die; and they, like us, vary in size and mood; in short, have individualities.

In structure, they are, like us, of symmetrical build and slightly polarized, although one cannot always tell the two ends easily apart, a feature perhaps correlated with their ability to move in opposite directions. Just how they move at all, we are unable to explain. What confounded us was that while most of them make contact with the ground at four points, others use only two; should the latter, more unstable, perhaps be rated as degenerative forms?

In volume, the majority fall roughly into three size classes. These evidently represent three different age groups, mass increasing with age, though not continuously, but rather in metamorphic steps. Moreover, our measurements of locomotor speeds

revealed a striking inverse relation between velocity and size, which would bear out the familiar rule that vigour and speed decline with age.

As we kept on observing, we could not fail to be struck by one peculiar constant attribute of Earthians: they had invariably associated with them some rather unimpressive bodies of much smaller size. These we have now definitely identified and classified as obligatory parasites of the Earthians, for the following good reasons:

1. They are lodged, for the most part, in the interior of Earthians.
2. Although at times they are disgorged to the outside, they never stray far off, and soon re-enter their hosts.
3. They are more numerous in larger hosts. Therefore, since host size reflects host age, the accessory bodies obviously multiply inside the hosts as the latter grow.
4. Even when detached from their hosts, the little bodies show only extremely limited capacity for independent active motion. Whatever motility we could observe contrasted drastically with that of Earthians by its sluggishness and, above all, its lack of direction.
5. Lastly, the fact that they make only unstable two-point contact with the ground likewise puts them into the degenerative class.

We have given these parasitic bodies the name of 'Miruses.'
Because of their plainly ancillary nature, we shall ignore them in
our detailed story, which now returns to our main subject—*The*
*Earthians* themselves. The evidence that these are truly living
creatures is overwhelming.

In the first place, they metabolize; that is, they take in substances from the environment, extract energy from them, and give off wastes, mostly as gas and smoke. Intake is mostly liquid, through two holes. One is in back; the other one in front. The front one seems to be much more important because it is quite near what

may be the brain, protected by a huge operculum. In fact, the two seem to have totally different functions. We judge this from a test made by a daring member of our party who managed surreptitiously to cross the feeder tubes. The Earthian so treated became completely paralyzed; his parasites, by contrast, showed signs of extreme agitation, the source of which remains unexplained.

Although most Earthians thus live on liquid food sucked in through tubes, the largest, hence oldest, of the race seem to have formed the habit of gulping solid matter with special organs sprouted for this purpose.

Some of the energy derived from intake goes into heat; and more than once, we saw it explode in spouts of vapours. The functional utility of these geysers is obscure. Some other parts of the energy, as noted earlier, is converted into light, and still another part, into sound. The volume of this sound grows as the square of population density.

Off and on, we noted signs of grooming by a wiping motion, but only in front. Astonishingly, this was started and stopped in synchrony by all the members of the population, as if they obeyed some secret commands. Does this imply that they do have brains?

Unquestionably, they do have an organ designed to integrate and shape data of information from the outside world into concerted actions—a sort of brain. For we found under their front

operculum a wild profusion of lines and tubes and links, so utterly mixed up and tangled that our 'systems analysts' have persuaded us of its basic resemblance to a thinking machine much like our own.

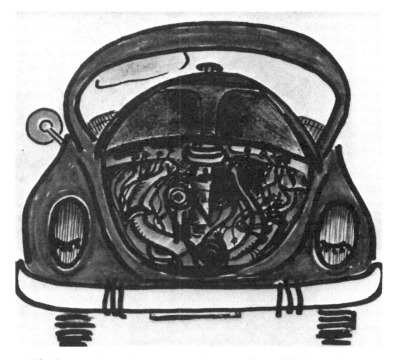

Life for the Earthian, ends as it must for all living beings, in death. Death either comes slowly, heralded by a phase of disarray, unsightly appearance, and frequent breakdowns, or suddenly in a violent noisy disintegration. As you will see, the fast kind has deep vital significance. Much as in our own world, the corpses are collected in heaps destined, at least in part, for some sort of reincarnation. In fact, it was from this source that we could gather clues as to the amazing method by which the Earthians propagate their kind. It took us an inordinate amount of time, effort, and imagination to reconstruct their mode of reproduction. Our conclusions, based partly on observation, partly on deduction, are these. The Earthians definitely do not reproduce by fission or budding. New Earthians are the product of true synthesis from elementary nonliving parts which are assembled stepwise in an orderly nonrandom sequence—a code. Life comes

into them only at the end of this process. The elementary constituents themselves are of uncertain origin. However, a series of
lucky incidents has made us favour the following theory:

On several occasions we noted two Earthians, fiercely attracted
to each other, embrace in a crushing hug, losing their shape,

vitality—indeed, identity—in the encounter. They evidently gave up their individual existence for a higher union. We could not help being mystified by this sacrificial act, till we discovered that it marked not an end, but rather a beginning: a renascence. In following the fate of the devitalized scraps picked up by other Earthians, we saw them vanish in a complex of huge structures which spouted fire and coloured smoke and later re-emerge as handsomely molded parts, preformed and ready for assembly into new Earthians. From this experience, we are convinced that the observed crash encounters are truly a mating of two individuals, which in loving abandon expunge their lives so that their merged substance may be reincarnated in living offspring. At times, a considerable lag period separates the act of love and the re-processing of its products to new life.

A rather curious correlation ought to be mentioned in this context. It was not at all uncommon for us to find pairs or groups of Earthians resting motionless in rather isolated spots. It could not escape us that, in the vicinity of such locations, the incidence of the described mating collisions was significantly higher. We wonder if this is sheer coincidence, or whether perhaps the observed phase of quiet togetherness might not be a sort of prelude to the mating crash.

We could go on documenting the living nature of these Earthians still further. However, we rest our case, having proven to everybody's satisfaction that they possess at least three of the basic attributes of life—metabolism, motility, and reproduction. That they are alive cannot be doubted. Being metallic, they might conceivably have come from the same primordial stuff as we ourselves. If so, they must have evolved at a much slower rate to have remained on such a primitive level of behaviour as we succeeded in recording. This gives us comfort; for it assures us that they could hardly ever come over to our lands to bother us. Far from disheartening, the discovery of life on earth has, on the contrary, strengthened our faith in our Martian supremacy as the unequalled climax of evolution.

Moreover, by actually witnessing the stepwise assembly of Earthians from nonliving scraps, our expedition has once more confirmed the brilliant deduction of our chemists, reached long ago, that *all* living matter can be synthesized from scratch. Not that we needed confirmation; for have we not *known* that life, wherever in the Universe it may exist, was not created, but has originated?

# The high energy physics colouring book

H J LIPKIN

From *Journal of Irreproducible Results* **12**, no 3 (1964).

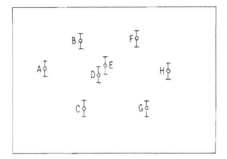

This is an experimental curve. Theory says there is a peak at point B. Colour the peak Red.

This is an experimental curve. Theory says there is no peak at point B. Colour the peak Grey.

This is an experimental curve. It is in complete disagreement with theory. Colour the error bars BLACK. Make them BIGGER, BIGGER!

This is a spark chamber picture An interaction at A produces three tracks: ABF, ACG, and ADEH. Draw in the tracks. Colour them any colours you wish and interpret the event.

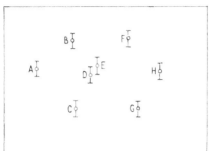

These are experimental points in a Feynman Diagram. Connect the points by appropriate solid, dashed, and wavy lines. Colour them in a gauge invariant way. Calculate the contribution of the diagram to all orders and disorders.

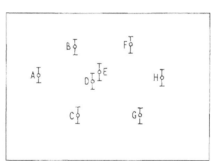

These points are experimental evidence for a new symmetry octet. There is NO time to colour this picture. Send it to *Phys. Rev. Letters* right away—or to the *New York Times*.

Can you find the Intermediate Boson in this picture?

The Poet, J. Alfred Neutrino
Who subsisted sublimely on vino,
With a spin of one-half
Wrote his own epitaph:
'No rest-mass, no charge, no bambino.

Jole Haag

# Snakes and Ladders

P J DUKE

From *Orbit*,
ournal of the
Rutherford High
Energy
Laboratory,
December 1963,
p 10.

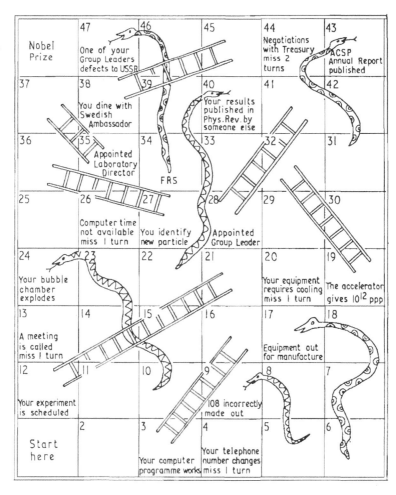

| 47 Nobel Prize | 46 One of your Group Leaders defects to USSR | 45 | 44 Negotiations with Treasury miss 2 turns | 43 ACSP Annual Report published |
|---|---|---|---|---|
| 37 | 38 You dine with Swedish Ambassador | 39 | 40 Your results published in Phys.Rev. by someone else | 41 | 42 |

(board game grid with the following labelled squares)

47 Nobel Prize
46 One of your Group Leaders defects to USSR
45
44 Negotiations with Treasury miss 2 turns
43 ACSP Annual Report published
37
38 You dine with Swedish Ambassador
39
40 Your results published in Phys.Rev. by someone else
41
42
36
35 Appointed Laboratory Director
34   FRS
33
32
31
25
26 Computer time not available miss 1 turn
27 You identify new particle
28 Appointed Group Leader
29
30
24 Your bubble chamber explodes
23
22
21
20 Your equipment requires cooling miss 1 turn
19 The accelerator gives $10^{12}$ ppp
13 A meeting is called miss 1 turn
14
15
16
17 Equipment out for manufacture
18
12 Your experiment is scheduled
11
10
9
8
7
Start here
2
3 Your computer programme works
4 Your telephone number changes miss 1 turn
5
6

---

## Do-it-yourself CERN Courier writing kit

From *CERN
Courier* 9
July 1969) p 211.

We present a 'writing kit' from which the reader himself may contruct a large variety of penetrating statements, such as he is accustomed to draw from our pages. It is based on the SIMP (Simplified Modular Prose) system developed in the Honeywell computer's jargon kit.

139

Take any four digit number—try 1969 for example—and compose your statement by selecting the corresponding phrases from the following tables (1 from Table A, 9 from Table B, etc. . .).

TABLE A

1. It has to be admitted that
2. As a consequence of inter-related factors.
3. Despite appearances to the contrary,
4. Until such time as fresh insight reverses the present trend,
5. Using the principle of cause and effect,
6. Presuming the validity of the present extrapolation,
7. Without wishing to open Pandora's box,
8. It is now proven beyond a shadow of a doubt that,
9. Worrying though the present situation may be,

TABLE B

1. willy-nilly determination to achieve success
2. construction of a high-energy accelerator
3. access to greater financial resources
4. pursuit of a Nobel prize
5. bubble chamber physics
6. a recent computation involving non semi-simple algebra
7. over-concern with the problems of administration
8. new measurements of eta zero zero
9. information presented in CERN COURIER

TABLE C

1. should only serve to add weight to
2. will inevitably lead to a refutation of
3. can yield conclusive information on
4. might usefully take issue with
5. must take into consideration
6. will sadly mean the end of
7. ought to stir up enthusiasm for
8. could result in a confirmation of
9. deflates the current thinking regarding

TABLE D

1. the need to acquire further computing capacity.
2. humanitarian concern with the personnel ceiling.
3. the Veneziano model.
4. a design which produces collisions at a later stage.
5. Macbeth's instruction 'Throw physic to the dogs'.
6. divergencies in weak interaction theory.
7. the desire to ensure that certain scientists go far.
8. bootstraps, conspiracies, poles and dips.
9. the future of physics in Europe.

# Gulliver's computer

JONATHAN SWIFT

From *Gulliver's Travels*, Part III 'A Voyage to Laputa' Chapter 5 (1727).

We crossed a Walk to the other Part of the Academy, where the Projectors in speculative Learning resided.

The first Professor I saw was in a very large Room, with Forty Pupils about him. After Salutation, observing me to look earnestly upon a Frame, which took up the greatest Part of both the Length and Breadth of the Room; he said, perhaps I might wonder to see him employed in a Project for improving speculative Knowledge by practical and mechanical Operations. But the World would soon be sensible of its Usefulness; and he flattered himself, that a more noble exalted Thought never sprang in any other Man's Head. Every one knew how laborious the usual Method is of attaining to Arts and Sciences; whereas by his Contrivance, the most ignorant Person at a reasonable Charge, and with a little bodily Labour, may write Books in Philosophy, Poetry, Politicks, Law, Mathematicks and Theology, without the least Assistance from Genius or Study. He then led me to the Frame, about the Sides whereof all his Pupils stood in Ranks. It was Twenty Foot square, placed in the Middle of the Room. The Superficies was composed of several Bits of Wood, about the Bigness of a Dye, but some larger than others. They were all linked together by slender Wires. These Bits of Wood were covered on every Square with Paper pasted on them; and on these Papers were written all the Words of their Language in their several Moods, Tenses, and Declensions, but without any Order. The Professor then desired me to observe, for he was going to set his Engine to work. The Pupils at his Command took each of them hold of an Iron Handle, whereof there were Forty fixed round the Edges of the Frame; and giving them a sudden Turn, the whole Disposition of the Words was entirely changed. He then commanded Six and Thirty of the Lads to read the several Lines softly as they appeared upon the Frame; and where they found three or four Words together that might make Part of a Sentence, they dictated to the four remaining Boys who were Scribes. This Work was repeated three or Four Times, and at every Turn the Engine was so contrived, that the Words shifted into new Places, as the square Bits of Wood moved upside down.

Six Hours a-Day the young Students were employed in this Labour; and the Professor shewed me several Volumes in large Folio already collected, of broken Sentences, which he intended to piece together; and out of those rich Materials to give the World a compleat Body of all Arts and Sciences; which however might be still improved, and much expedited, if the Publick would

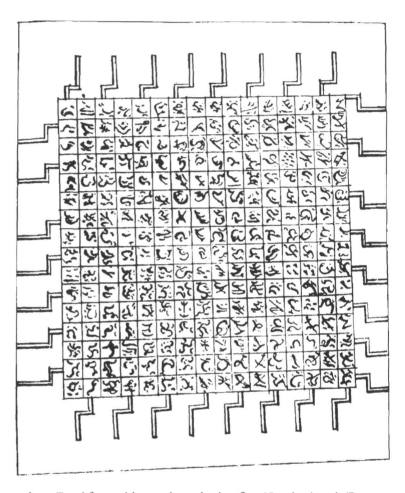

raise a Fund for making and employing five Hundred such Frames in *Lagado*, and oblige the Managers to contribute in common their several Collections.

He assured me, that this Invention had employed all his Thoughts from his Youth; that he had emptyed the whole Vocabulary into his Frame, and made the strictest Computation of the general Proportion there is in Books between the Numbers of Particles, Nouns, and Verbs, and other Parts of Speech.

# Haiku

Adapted from the catalogue of the exhibition *Cybernetic Serendipity—the Computer and the Arts*, Studio International, London 1968 p 53; and *NPL News* 204, 10 (1967).

The haiku form is simple—a verse of 17 syllables, divided into three lines of five, seven and five syllables respectively. The Western ear should note that the metrical unit is the syllable (Japanese is a syllabic language) and not, as in Western prosody, the foot composed of one or two syllables. The form of 17 syllables is not chance; it derives from the traditional view of Japanese linguistic philosophy that 17 syllables is the optimum length of human speech to be delivered clearly and coherently in one breathing.

These examples were produced by on-line man-machine interaction at the Cambridge Language Research Unit. The programme provides a frame with 'slots' in which the operator types words. His choice is constrained by the lists and arrow directions in the thesaurus and diagram. These show that the semantic centre of the poem—with five arrows going to it and one going from it—is situated at slot 5.

## THESAURUS

| Slot 1 (→4) (→5) | Slot 2 (→5) (→6) | Slot 3 (→5) | Slot 4 (→6) (→7) | Slot 5 (→8) | Slot 6 (→2) | Slot 7 | Slot 8 (→5) | Slot 9 (→5) (→8) |
|---|---|---|---|---|---|---|---|---|
| White | Buds | See | Snow | Trees | Spring | Bang | Sun | Flit |
| Blue | Twigs | Trace | Tall | Peaks | Full | Hush | Moon | Fled |
| Red | Leaves | Glimpse | Pale | Hills | Cold | Swish | Star | Dimmed |
| Black | Hills | Flash | Dark | Streams | Heat | Pffftt | Cloud | Cracked |
| Grey | Peaks | Smell | Faint | Birds | Sun | Whizz | Storm | Passed |
| Green | Snow | Taste | White | Specks | Shade | Flick | Streak | Shrunk |
| Brown | Ice | Hear | Clear | Arcs | Dawn | Shoo | Tree | Smashed |
| Bright | Sun | Seize | Red | Grass | Dusk | Grrr | Flower | Blown |
| Pure | Rain | | Blue | Stems | Day | Whirr | Bud | Sprung |
| Curved | Cloud | | Green | Sheep | Night | Look | Leaf | Crashed |
| Crowned | Sky | | Grey | Cows | Mist | Crash | Child | Gone |
| Starred | Dawn | | Black | Deer | Trees | | Crane | Fogged |
| | Dusk | | Round | Stars | Woods | | Bird | Burst |
| | Mist | | Square | Clouds | Hills | | Plane | |
| | Fog | | Straight | Flowers | Pools | | Moth | |
| | Spring | | Curved | Buds | | | | |
| | Heat | | Slim | Leaves | | | | |
| | Cold | | Fat | Trees | | | | |
| | | | Burst | Pools | | | | |
| | | | Thin | Drops | | | | |
| | | | Bright | Stones | | | | |
| | | | | Bells | | | | |
| | | | | Trails | | | | |

SEMANTIC SCHEMA

ALL ..... IN THE .....
(1) (2)

I ....... ....... ...... IN THE .....
(3) (4) (5) (6)

..... THE ..... HAS .....
(7) (8) (9)

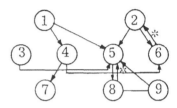

An asterisk indicates a double linkage. For the system to be computable only one arrow must be chosen.

THREE RESULTS

*All green in the leaves*
*I smell dark pools in the trees*
*Crash the moon has fled*

*All white in the buds*
*I flash snow peaks in the spring*
*Bang the sun has fogged*

*All starred in the cold*
*I sieze thin trails in the mist*
*Look the moth has gone*

Here are two haiku written by human members of the NPL

*Pattern perception*
*Is easier to do than*
*Cerebrate about*

*Don't design systems*
*Of automatic control*
*Ride a bicycle*

---

## Textbook selection

MALCOLM JOHNSON

Publishers note the tendency of some teachers to consider prestige of a title above the needs of their particular students in the selection of a textbook. A former editor of one of the largest publishers of technical books tells of this incident.

'At our main editorial meeting I was describing a new and important project at Caltech for which I wanted to offer a contract, and stated that the title of the book was to be *Elementary Particle Theory*. Hearing this, the Vice President in charge of our International Division, ever mindful of the foreign demand for high level books, then told me that if I could suggest that the authors remove the word 'Elementary' from the title, he would be able to sell another thousand copies in the international market.'

# Computer, B.Sc. (failed)

E MENDOZA

Adapted from the catalogue of an exhibition *Cybernetic Serendipity— the Computer and the Arts* Studio International, London, 1968 p 58.

In 1960, physics students were still asked to write essays in English, but this procedure was not without its critics. A senior member of the Physics Department at Manchester once remarked—in a moment of cynicism—that he did not think that students ever really learned the concepts of physics. All they learned (he said) was a vocabulary of okay words for essay-writing, which they strung together in arbitrary order, relying on the fact that an examiner pressed for time would not actually read what they had written but would scan down the pages looking for these words. He did not mean to be taken very seriously but it struck me that we could test this hypothesis.

I produced a computer program—novel for those days—which was an elaborate version of the 'CERN Courier Kit' (page 139). There were alternative paths between the groups of words and quite complicated grammatical constructions could be incorporated, but the associations of phrase with phrase were pretty random and indeed I called them 'High Entropy Essays'. Eventually, I got in touch with a colleague at another university. About a week before a first-year exam, he sent me some question papers together with a suitable false name and one of the special books for writing the answers in; it was all done very secretly. To answer the questions, I wrote suitable random vocabularies. The computer churned out its nonsense and I copied it into the book . . . The script was eventually slipped in among the genuine ones. In fact, it was marked by a conscientious examiner who read every word and stormed into the Director's office shouting 'Who the hell is this man ? Why did we ever admit him ?' So the hypothesis about students' learning was proved incorrect.

Here is one question and answer. Three different vocabularies about the standard meter and interferometry were used.

QUESTION: Describe one method by which the wavelength of a spectral line has been compared with the standard metre.

ANSWER: The standard metre which since 1700 has been determined by accurate instruments is a lump of metal at the Quai d'Orsay. The substandard metre by agreement has been determined in spite of difficulties and by agreement the unit length is a metal mass of low expansion coefficient. The position of the scratches can be measured microscopically. The dispersion can be correlated with multiple reflections and the fringe width is optimized with large apertures. The resolving power which increases usually with a coherent source is limited by the ratio d $\lambda/\lambda$ while the dispersion which can be correlated with the ratio d $\lambda/\lambda$ sometimes depends on the number of lines per inch. In all cases the resolving power is determined by the number of lines per inch. The fringe width becomes maximized with long optical path differences. The optical constants determine the flexibility. The path length enters as in the vector diagram. Nowadays the fringe width enters, as in Newton's rings and following Michelson the fringe width plays an essential role. It limits the stability as a standard while the monochromaticity obviously defines the need for good contrast.

## Collective names in basic sciences

ANON

From *Journal of Irreproducible Results* **14**, 4 (1965).

A pile of nuclear physicists
A grid of electrical engineers
A set of pure mathematicians
A field of theoretical physicists
An amalgamation of metallurgists
A line of spectroscopists
A coagulation of colloid chemists
A galaxy of cosmologists
A cloud of theoretical meteorologists
A shower of applied meteorologists
A litter of geneticists
A knot of nautical engineers
A labyrinth of communication engineers
An exhibition of Nobel prize winners
An intrigue of council members
A dissonance of faculty members
A stack of librarians
A chain of security officers
A complex of psychologists
A wing of ornithologists
A batch of fermentation chemists
A colony of bacteriologists

# The Chaostron. An important advance in learning machines

J B CADWALLADER-COHEN, W W ZYSICZK and
R R DONELLY

Condensed from
*Journal of Irreproducible Results*
**10**, 30 (1961).

[*One of the most interesting directions of computer research is to see whether a computer can be made to* learn *to solve problems. This account of some early experiments—reported in 1961 from the Bell Telephone Laboratories although apparently none of the authors were known employees of that organization—started from the observation that animals, when set a mechanical problem to solve, usually begin by performing purely random actions for a considerable time and then find the solution quite by accident. This seems to be their learning process. This provided the basis for the Chaostron experiments.*]

The authors feel strongly that the key to successful automation of learning tasks lies in randomization of the response patterns of the machine. The failure of various previous attempts in this direction, we feel, has been due to two problems: first, the difficulty of getting a sufficient degree of randomness built into the structure of the machine, and second, the expense of creating a device large enough to exhibit behaviour not significantly influenced by the operation of any one of its components. We are deeply indebted to Dr R Morgan for a suggestion which showed us the way out of these difficulties: design for the Chaostron was done by taking 14 000 Western Electric wiring charts, cutting them into two-inch squares, and having them thoroughly shaken up in a large sack, then glued into sheets of appropriate size by a blindfolded worker. Careful checks were made during this process, and statistical tests were made on its output to insure against the propagation of unsuspected regularities.

*Simulation experiments*

Unfortunately, we have not, as yet, been able to complete the wiring of Chaostron. We felt, however, that it should be possible to estimate the effectiveness of Chaostron even before its completion by simulating it on a high speed digital computer. This procedure had the further advantage of attracting the interest of representatives of the Bureau of Supplies and Accounts of the US Navy, who found in Chaostron an excellent aid to control of the Navy's spare parts inventory. The Navy, as a result, was generous enough to offer time on a BuShips computer for the simulation of Chaostron.

The computer of choice for the simulation runs was the IBM STRETCH machine which not only operates at very high speed, but is also able to accept input programs coded in YAWN language, which closely resembles colloquial English. We felt it very important to use a source language for simulation programs which would contain as much ambiguity as ordinary speech, since undue precision in the simulation programs might accidentally 'tip off' the machine as to the nature of the desired solutions.

In the event, it was not possible to obtain a STRETCH computer for the project, and so simulation was done by simulating STRETCH simulating Chaostron on an IBM 704. All simulation runs were conducted in essentially similar universes of environments; the computer was presented with a sequence of circles, squares and crosses represented by punched cards, and was required to print, after examining each stimulus, one of the words 'circle,' 'square' or 'cross'. No reinforcement from the experimenter was provided, since it was feared that such reinforcement would bias the learning process, and thus vitiate the validity of any conclusions we might wish to draw from the results.

The first trials were run with the input stimuli represented on the punched cards as geometric patterns of punching in the appropriate shapes. As a control, one run was made with no stored program initially in the machine to check that the learning rate of the untutored machine was not so great as to interfere with further studies. For this run, the machine memory was cleared, the cards containing patterns were placed in the card reader, and the load card button was pressed. After three hours the machine had not printed its response to the first input pattern; evidently the rate of learning under these conditions is very low (we judge it to be of the order of $10^{-6}$ concepts per megayear).

Therefore we proceeded with the main series of experiments, in which a random program was loaded into the computer ahead of each batch of data cards. A total of 133 random programs were tried in random sequence. Even in this series of experiments the machine took a surprisingly long time to respond to the stimuli; in most cases the run had to be terminated before the first response occurred. However, on run number 73, the computer responded

*** / AX$, )U,,,, .

to the first stimulus card (which was a square); on run 114 the computer responded

66666666666666666666666666666666666

to every stimulus; and on run 131 the computer ejected the printer paper twice.

*Conclusions*

Unfortunately, budget difficulties forced us to abandon this approach after 133 trials, in spite of the promising appearance of the early results. Thus, our conclusions are perforce based on a smaller data sample than we would like. Nonetheless, certain points are clear.

1. The correlative reinforcement model of learning advanced by Dewlap *et al* is untenable in view of our results. No triphasic system could function without a degree of organization exceeding that which we have used in the simulation studies. Even this degree of structure, however, resulted in extremely slow response to comparatively simple stimuli.

2. It seems evident that further understanding of machine learning requires resynthesis in operational terms of the conceptual framework provided by the Liebwald–Schurstein–Higgins suggestion that memory traces are renewed by associative stochastic increments to ideometric pathways shared by stimulus-coupled functional elements.

3. Not only is machine learning possible, but in fact it occurs under conditions of considerable difficulty. Indeed, it appears that even the simplest machines have a great amount of innate 'curiosity' (where by 'curiosity', of course, we do not mean to imply that anthropomorphic categories or judgments should be applied to machines, but merely that the machines have a desire to learn).

Our acknowledgements and thanks are due to Mr J B Puffader, for his assistance in the detailed design of Chaostron, and to Mr V A Vyssotsky for manually simulating the 704 simulating STRETCH simulating Chaostron, to complete run 133 after the budget funds ran out.

---

From *The Dry Rot of our Academic Biology*, by W M Wheeler, *Science* 57 pp 61–71 (1923). Many of us coddle our graduate students till the more impressionable of them develop the most sodden types of the father-complex. Some of us even wear out a layer of cortical neurones annually, correcting their spelling and syntax. One fussy old guru of my acquaintance has destroyed both of his hemispheres, his corpus callosum and a large part of his basal ganglia hunting stray commas, semicolons, dashes, parentheses and other vermin in doctors' dissertations.

# Physics is too young

From *Philosophy of the Inductive Sciences* vol 2, book 13 (1847).

[*William Whewell was an active scientist, one of the first philosophers of science and an enlightened reformer of university education at Cambridge. Nevertheless he divided the sciences into permanent and progressive ones, to the detriment of the latter. Mathematics and Newton's mechanics were permanent while new-fangled geology and chemistry were liable to reinterpretation* ('There is nothing old, nothing stable'). *They should therefore be excluded from the University curriculum. Whewell was saying this in 1847; as a result the first professor of physics was not appointed in Cambridge till 1874.*]

We may assert in general that no Ideas are suited to become the elements of elementary education, till they have not only become perfectly distinct and fixed in the minds of the leading cultivators of the science to which they belong; but till they have been so for some considerable period. The entire clearness and steadiness of view which is essential to sound science, must have time to extend itself to a wide circle of disciples. The views and principles which are detected by the most profound and acute philosophers, are soon appropriated by all the most intelligent and active minds of their own and of the following generations; and when this has taken place (and not till then), it is right, by a proper constitution of our liberal education, to extend a general knowledge of such principles to all cultivated persons. And it follows, from this view of the matter, that we are by no means to be in haste to adopt, into our course of education, all new discoveries as soon as they are made. They require some time, in order to settle into their proper place and position in men's minds, and to show themselves under their true aspects; and till this is done, we confuse and disturb, rather than enlighten and unfold, the ideas of learners, by introducing the discoveries into our elementary instruction. Hence it was perhaps reasonable that a century should elapse from the time of Galileo before the rigorous teaching of mechanics became a general element of intellectual training; and the doctrine of universal gravitation was hardly ripe for such an employment till the end of the last century. We must not direct the unformed youthful mind to launch its little bark upon the waters of speculation, till all the agitation of discovery, with its consequent fluctuation and controversy, has well subsided.

# Yes, Virginia

Adapted from
*American Journal
of Physics* 33,
345 (1965).

[*Every American knows that in 1897 a young girl wrote a letter to the
Editor of the New York Sun:*

*Dear Editor:*
*I am 8 years old.*
*Some of my little friends say there is no Santa Claus.*
*Papa says 'If you see it in "The Sun" it's so.'*
*Please tell me the truth, is there a Santa Claus?*
VIRGINIA O'HANLON,
*115 West 95th Street, New York City.*

*The Editor replied: 'Yes Virginia, there is a Santa Claus . . . and the
letters have been reprinted each Christmas season ever since.*

*In 1964, Vernet E Eaton of Wesleyan University spoke on 'The Demon-
stration Lecture as an Art'. The lecturer had apparently received a letter
from Virginia's granddaughter (also called Virginia) and had replied
to it.*]

Dear Professor Eaton:
I am 18 years old and a Freshman at Marsupial State College. I am
disturbed about our physics course. Although we have finished
mechanics, so far we have had no demonstrations and have seen
no apparatus.

My classmates are not disturbed. They boast that when they
grew up they quit playing with toys and that I should not cry
when my toys are taken away. They claim that acceleration is a
second derivative and we should not confuse the issue by bringing in carts and falling bodies. According to them angular
momentum is a vector and we should not worry about turning
wheels.

My father is a loyal Wesleyan man and believes you are always
right. He has agreed that if you say that my classmates are wrong
I may transfer to Empirica University where demonstrations play
an important part in the learning process.
Very truly yours,
VIRGINIA REEHL

Dear Virginia:
Your classmates are wrong. Although they think they are being
modern, it seems to me that they are living in ancient Greece
where manual work and manipulation was delegated to slaves
while free men talked and figured.

No matter how convincing the proof, no theory becomes a law
until it has been tested by experiment. Have you read Aristotle's

convincing proof that heavy bodies fall faster than light bodies? This illustrates how easy it is to prove something that isn't true and of course Galileo showed how a simple experiment may point out the fallacy. Even Maxwell's brilliant work was not accepted until Hertz proved in the laboratory that electromagnetic waves do exist. Remind your classmates that Maxwell provided the theory and mathematical rigour but Hertz put us in the driver's seat.

Some seem to feel that once they have written the Hamiltonian the job is done. I suspect they imagine that eating consists of consuming differential equations and they go on a diet by changing the upper limit of the integral. Some go so far as to claim that since the results are based primarily on shape and curvature, beauty contests should be decided on the basis of the constants in a Fourier series. Please do not conclude that mathematical rigour and sound logic are not important. They are extremely important. Mathematics is a necessary but not sufficient tool for understanding nature.

Would this not be a dreary world without apparatus to talk to us. Like music and art and nature, however, it speaks to us in a language without words and not everybody understands this language.

I will be glad to recommend you to Empirica U.

Very truly yours,

VERNET E EATON

Wesleyan University, Middletown, Connecticut

---

From *Physicists continue to laugh* MIR Publishing House, Moscow 1968. Translated from the Russian by Mrs Lorraine T Kapitanoff.

Once in Russia, in a physics exam, the professor wrote the equation

$$E = h\nu$$

and asked a student:

'What is $\nu$?'
'Planck's constant.'
'And $h$?'
'The length of the plank.'

[*Astonishingly, this is translated directly from the Russian.*]

# How to learn

LEWIS CARROLL

From *The Complete Works of Lewis Carroll* (London: The Nonesuch Press) 1939 pp 1116–19.

[*The Lorentz contraction and spatial transforms seem to be detectable in Lewis Carroll's 'Through the Looking Glass'. Reciprocally, when the Rev. Charles Lutwidge Dodgson wrote on mathematics and logic it was not without whimsy, as evidenced by the following. In any case, it is excellent advice on how to read a textbook.*]

The learner, who wishes to try the question *fairly*, whether this little book does, or does not, supply the materials for a most interesting mental recreation, is *earnestly* advised to adopt the following Rules:

1. Begin at the *beginning*, and do not allow yourself to gratify a mere idle curiosity by dipping into the book, here and there. This would very likely lead to your throwing it aside, with the remark 'This is *much* too hard for me!', and thus losing the chance of adding a very *large* item to your stock of mental delights . . .

2. Don't begin any fresh Chapter, or Section, until you are certain that you *thoroughly* understand the whole book *up to that point*, and that you have worked, correctly, most if not all of the examples which have been set . . . Otherwise, you will find your state of puzzlement get worse and worse as you proceed, till you give up the whole thing in utter disgust.

3. When you come to a passage you don't understand, *read it again*: if you *still* don't understand it, *read it again*: if you fail, even after *three* readings, very likely your brain is getting a little tired. In that case, put the book away, and take to other occupations, and next day, when you come to it fresh, you will very likely find that it is *quite* easy.

4. If possible, find some genial friend, who will read the book along with you, and will talk over the difficulties with you. *Talking* is a wonderful smoother-over of difficulties. When *I* come upon anything—in Logic or in any other hard subject—that entirely puzzles me, I find it a capital plan to talk it over, *aloud*, even when I am all alone. One can explain things so *clearly* to one's self! And then, you know, one is so *patient* with one's self: one *never* gets irritated at one's own stupidity!

If, dear Reader, you will faithfully observe these Rules, and so give my little book a really *fair* trial, I promise you, most confidently, that you will find Symbolic Logic to be one of the most, if not *the* most, fascinating of mental recreations! In this First Part, I have carefully avoided all difficulties which seemed to me to be beyond the grasp of an intelligent child of (say) twelve or fourteen years of age. I have myself taught most of its contents, *viva voce*, to *many* children, and have found them take a real intelligent interest in the subject. For those, who succeeded in mastering Part I, and who begin, like Oliver, 'asking for more,' I hope to provide, in Part II, some *tolerably* hard nuts to crack—nuts that will require all the nut-crackers they happen to possess!

Mental recreation is a thing that we all of us need for our mental health. Symbolic Logic will give you clearness of thought—the ability to *see your way* through a puzzle—the habit of arranging your ideas in an orderly and get-at-able form—and, more valuable than all, the power to detect *fallacies*, and to tear to pieces the flimsy illogical arguments, which you will continually encounter in books, in newspapers, in speeches, and even in sermons, and which so easily delude those who have never taken the trouble to master this fascinating Art. *Try it.* That is all I ask of you!

## The nature of evidence

ISAAC TODHUNTER

From *The Conflict of Studies and Other Essays* (Macmillan) 1873.

[*Writing in 1873, the mathematician Isaac Todhunter had a poor opinion of Experimental Philosophy, which was then being introduced into Cambridge teaching. In particular he despised practical classes.*]

We assert that if the resistance of the air be withdrawn a sovereign and a feather will fall through equal spaces in equal times. Very great credit is due to the person who first imagined the well-known experiment to illustrate this; but it is not obvious what is the special benefit now gained by seeing a lecturer repeat the process. It may be said that a boy takes more interest in the matter by seeing for himself, or by performing for himself, that is by working the handle of the air-pump: this we admit, while we continue to doubt the educational value of the transaction. The boy would also probably take much more interest in foot-ball than in Latin grammar; but the measure of his interest is not identical with that of the importance of the subjects. It may be said that the fact makes a stronger impression on the boy through the medium of his sight, that he believes it the more confidently. I say that this ought not to be the case. If he does not believe the statements of his tutor—probably a clergyman of mature knowledge, recognized ability, and blameless character—his suspicion is irrational, and manifests a want of the power of appreciating evidence, a want fatal to his success in that branch of science which he is supposed to be cultivating.

# School leaving exam

[*For the first time in 1858, Oxford and Cambridge Universities conducted public examinations, called School Leaving Examinations for junior and senior candidates, which could be taken at a number of local centres. From these originals are descended today's Ordinary and Advanced levels of the General Certificate of Education—the two hurdles which almost every English schoolboy physicist must surmount.*

*In the very first 'Oxford local' junior examination, there was a paper on Mechanics and the Mechanism of the Steam Engine: a total of 29 questions of which the candidates were 'expected to attempt not more than ten'. Here are some.*]

9. In a screw press, the screw has 4 threads to the inch, the power is applied at a distance of 14 inches from the axis of the screw, and the surface pressed is 110 square inches. Find (approximately) what power must be applied to produce a pressure of 1 lb to the square inch?

13. A bow is stretched until the tension of the string just equals the pulling force. What is the angle between the two parts of the string?

21. What causes the *puffing* noise of a locomotive engine? If 4 puffs be heard in a second, and the circumference of the driving wheel be 22 feet, how many miles an hour is the train going?

[*On the Natural Philosophy paper of the senior Examination, all the questions were descriptive with no calculations. There were 15 on the paper, candidates being 'recommended not to attempt many of the questions, but to select a few which they may be able to answer correctly.' Here are two.*]

3. What is said to be the velocity of light? How was it first deduced by Römer, and subsequently confirmed by Bradley? Explain fully their observations, and the deductions from these observations. How far would they shew the similarity of the nature of light when derived from different sources? Do you know in what countries and at what date Römer and Bradley lived? Do all kinds of sounds travel with equal velocities? Prove your answer, and mention any facts you know as to the velocity with which they traverse different media. How did Wheatstone try to deduce the velocity of electricity? How much did his experiment really prove? What circumstances practically affect the velocity of electricity, and at what rate is it found to travel in practice? Does its velocity depend upon the intensity of the battery from which it is produced? Compare the laws which seem to regulate the velocities of light, sound, and electricity.

12. What is the distinction between the momentum of a body and its '*vis viva*?' Supposing an anvil or large stone to be placed on a man's chest, explain fully all the reasons why a blow with a hammer on the stone or anvil will do him no injury.

---

## Where to hold nuclear spectroscopy conferences in Russia

From the ITEP newspaper *People and Spectra*, 1965.

In the immediate future it will prove more and more difficult to select places for holding meetings. Before giving some kind of advice about this we would like to get one thing clear: are these meetings arranged with lofty scientific aims or are they purely for entertainment? At present this is not obvious to us. If it turns out that we meet annually so as to enjoy these ten days, then let us abolish those boring reports and tiresome excursions and transfer the meetings to summertime.

But if it appears to us that the meetings are after all for science then it is necessary to organize them differently so that nothing will distract the participants from their main task. In this case it is possible to recommend the following places for holding meetings.
1. *Congress cave in the Urals*. To conduct meetings on nuclear spectroscopy in caves is very useful since the background of ionizing radiation is very low. Congress cave in this respect is especially good since it is icy and the natural level of radioactivity in it must be lower than usual. The constant low temperature in the cave will promote liveliness in the participants and completely exclude sleepiness even during review reports.
2. *The valley of geysers on Kamchatka*. A very warm little place. Unusual phenomena of nature—geysers—will evoke in the participants the desire to speculate on various scientific themes, perhaps including nuclear spectroscopy. The ground here trembles a little all the time which will also hinder sleep during the sessions. At the same time it is possible to solve the problem of keeping to time, by placing the reporter's lectern over a geyser which operates regularly every five minutes.
3. *The atomic ice-breaker 'Lenin'*. This is unnecessary to describe; it is clear that nuclear spectroscopy has a direct relationship to it. It is advisable to hire the ice breaker at a time after it has broken all the ice in the Arctic and is heading towards the Southern Hemisphere.

# Typical examination questions as a guide to graduate students studying for prelims

H J LIPKIN

From *Journal of Irreproducible Results* 7, 12 (1958)

## 1. MECHANICS

A particle moves in the potential well $V(r) = e^{-r}/r^{12}$

(a) Show that the solution of this problem tells nothing about the binding energy of the deuteron.

(b) Discuss the asymptotic behaviour of the solution as $12 \to \infty$

## 2. ELEMENTARY PARTICLES

List all elementary particles which have not yet been discovered, giving mass, charge, spin, isotopic spin, strangeness, and the reason why they have not yet been observed. Discuss the changes which will be necessary in current theories as each one is discovered.

## 3. GENERAL RELATIVITY

A clock is placed on a rotating table. As the speed of rotation is increased, the clock flies to pieces. Discuss the angular distribution of the fragments and show how this allows one to calculate the spin and parity of the clock.

## 4. QUANTUM THEORY

(a) Write the Schrödinger Equation for an undergraduate taking Elementary Physics. Develop the appropriate operators for the description of the system, including the pass–flunk operator which has the eigenvalue $+1$ if the student passes the course and $-1$ if he flunks. Show that the student's state at the end of the term is always an eigenstate of this operator.

(b) Show that if all undergraduates were transformed into Hilbert Space it would be a good thing.

## 5. INVARIANCE PROPERTIES

(a) Discuss the behaviour under space inversion, charge conjugation and time reversal of the Schrödinger equation $H\Phi = E\Phi$. Note that under space inversion this equation becomes $\Phi \exists = \Phi H$. Discuss the implications of this property in proving the existence of solutions of the Schrödinger equation.

(b) Discuss the behaviour of the Dirac Equation under rotation:

(i) when the blackboard upon which the equation is written is rotated;

(ii) when the physicist studying the equation is rotated.

## 6. NUCLEAR REACTIONS

A nickeleon enters a cokemachine nucleus. Discuss the relative probabilities of the following reactions:

(*a*) An (n,c) reaction ( = nickeleon in, cocacolon out)

(*b*) Absorption ( = nickeleon in, nothing out)

(*c*) Elastic scattering (n,n) ( = nickeleon in, nickeleon out)

(*d*) An (n,2n) reaction ( = nickeleon in, two nickeleons out)

(*e*) An (n,p) reaction ( = nickeleon in, penniton out)

(*f*) An (n,d,) reaction ( = nickeleon in, dimeteron out)

Consider also the effect of selection rules, depending upon the spin of the nickeleon, and of perturbations, such as banging or tilting the machine.

## 7. RELATIVISTIC QUANTUM FIELD THEORY

A pair of twins, named Bingle and Dingle, are separated at birth, and Dingle is sent off to a distant star, at a velocity of $0.999c$ and returns. Discuss the relative ages of Bingle and Dingle, taking into account the following effects:

(*a*) Bingle and Dingle exchange light signals continuously during the trip

(*b*) At the turning point (vortex) of Bingle's journey, he emits a virtual pion which then creates a Dingle–Anti-Dingle Pair. The Anti-Dingle returns to earth and annihilates the original Dingle, while the remaining Bingle and Dingle remain on the distant star and live happily ever after.

## 8. EXPERIMENTAL TECHNIQUES

A beam of optically-pumped polarized rubidium atoms is passed through a homogeneous magnetic field and a radio frequency field. This is followed by a passage through a thin foil of magnetized iron and an adiabatic fast passage through an inhomogeneous electric field, two mutually perpendicular gravitational fields, a radio frequency scalar meson field, and a wheat field.

(*a*) Why?

(*b*) Describe an experiment to measure Planck's constant using the most expensive equipment possible.

## 9. THE MANY-BODY PROBLEM

Discuss the properties of a system of strongly interacting physicists coupled to a high energy accelerator. Give particular attention to the following points:

(*a*) The independent physicist model (stay-in-the-shell-model)
(*b*) The collective model
(*c*) Pairing correlations and quasi-physicists
(*d*) Linked clusters
(*e*) Seniority
(*f*) Fractional Parentage
(*g*) Acceleration in real and virtual states

10. SUPERCONDUCTIVITY

Explain the relativistic East–West effect in superconductivity. Show by the use of the Leningrangian formalism that a transformation exists by which an Eastern theory which has occurred later than a Western theory can be made to antedate the Western theory.

11. DISPERSION RELATIONS

Explain the multiple production of strange articles in unclear physics appearing on the non-physical region of the *Physical Review*. Show that the principle of causality allows the complete prediction of results which are in good agreement with experiment until the experiment is performed. Using the formal theory of scattering, discuss the validity of (*a*) the impulse approximation, (*b*) the Born approximation, (*c*) the Unborn approximation. Show that clarity of the *Physical Review* increases exponentially with the number of articles which are left in the Unborn approximation.

---

## Big Science and Lesser Sciences

P M S BLACKETT

From '*Memories of Rutherford*' in *Rutherford at Manchester* ed J B Birks (London: Heywood) 1962.

His prestige was such that even a joke from Rutherford's mouth was apt to become a dogma in lesser men's minds. No very young physicist could be totally unaffected by his famous crack: 'All science is either physics or stamp collecting', or by the often implied assumption that it only needed some further progress in physics to allow us to deduce from first and physical principles the facts and laws of the lesser sciences like chemistry.

# Oral examination procedure

S D MASON

From *Proceedings of the IRE*, May 1956 p 696.

In these brief notes the purposes of an oral examination are set forth and practical rules for conducting one are given. Careful attention to the elementary rules is necessary in order to assure a truly successful examination. From the standpoint of each individual examiner the basic purposes of the oral examination are: to make that examiner appear smarter and trickier than either the examinee or the other examiners, thereby preserving his self esteem, and to crush the examinee, thereby avoiding the messy and time-wasting problem of post-examination judgment and decision.

Both of these aims can be realized through diligent application of the following time-tested rules:

1. Before beginning the examination, make it clear to the examinee that his whole professional career may turn on his performance. Stress the importance and formality of the occasion. Put him in his proper place at the outset.

2. Throw out your hardest question first. (This is very important. If your first question is sufficiently difficult or involved, he will be too rattled to answer subsequent questions, no matter how simple they may be.)

3. Be reserved and stern in addressing the examinee. For contrast, be very jolly with the other examiners. A very effective device is to make humorous comments to the other examiners about the examinee's performance, comments which tend to exclude him and set him apart, as though he were not present in the room.

4. Make him answer each problem your way, especially if your way is esoteric. Constrain him. Impose many limitations and qualifications in each question. The idea is to complicate an otherwise simple problem.

5. Force him into a trivial error and then let him puzzle over it for as long as possible. Just after he sees his mistake but just before he has a chance to explain it, correct him yourself, disdainfully. This takes real perception and timing, which can only be acquired with some practice.

6. When he finds himself deep in a hole, never lead him out. Instead, sigh, and shift to a new subject.

7. Ask him snide questions, such as, 'Didn't you learn that in Freshman Calculus?'

8. Do not permit him to ask you clarifying questions. Never repeat

or clarify your own statement of the problem. Tell him not to think out loud, what you want is the answer.

9. Every few minutes, ask him if he is nervous.

10. Station yourself and the other examiners so that the examinee can not really face all of you at once. This enables you to bracket him with a sort of binaural crossfire. Wait until he turns away from you toward someone else, and then ask him a short direct question. With proper coordination among the examiners it is possible under favorable conditions to spin the examinee through several complete revolutions. This has the same general effect as item 2 above.

11. Wear dark glasses. Inscrutability is unnerving.

12. Terminate the examination by telling the examinee, 'Don't call us; we will call you.'

### Fluorescent yield

ARTHUR H SNELL

*There was an electron in gold*
*Who said, 'Shall I do as I'm told?*
*Shall I snuggle down tight*
*With a brief flash of light*
*Or be Auger outside in the cold?'*

But there are many possibilities and equally many poems. On internal conversion, for instance:

*Said a K-shell electron in gold,*
*'I m thinking of leaving the fold*
*To be hit like a hammer*
*By an outgoing gamma.*
*In freedom I'll live till I'm old.*

Or even on electron capture:

*Said the K-shell electron in gold,*
*'I wonder if I might be bold,*
*And make a slight shift*
*From this circular drift*
*And change this damned atom to platinum.'*

161

# Slidesmanship

D H WILKINSON

Condensed from 'Elements of Conferencemanship,' Proceedings of the International Conference on Nuclear Structure, Kingston, Canada. (North-Holland and University of Toronto Press 1960) pp 906–12.

My present communication is torn from its proper context of Conferencemanship of which it is merely one, and not the most important, of the many facets. I shall have no opportunity to enlarge upon, for example, 'How to mention your collaborators without actually giving them any credit' or upon 'How to discredit your rival's theory and experimental technique without understanding either.'

Slidesmanship has three main divisions. Of the third, 'The subjugation of your personal adversary' I am not permitted to speak. The other two are 'The subjugation of the projectionist' and 'The subjugation of the audience.'

It is the Slidesman's task to wrest the apparent initiative from the projectionist and to reduce him to a nervous pulp. It is important for the Slidesman to know when he has succeeded, because only then can he turn his full attention to his audience, which is, after all, his major task. Since the projectionist is usually invisible it is a little tricky to be sure when he has been pulped but I myself find it quite satisfactory to continue until his gibbering is clearly audible.

I do not recommend to any but the veriest tyros crude and vulgar techniques such as the intimate interleaving of 35 mm and regular size slides, or even the use of the once-popular pentagonal slide. A satisfactory beginning for the more aspiring is the '3-2-1' technique. It exploits the fact that the projectionist always loads up the first two slides when the chairman announces the talk so he can snap one on to the screen as soon as the speaker says 'First . . .' and follow like a machine gun with the second if need be. The Slidesman therefore begins: 'Third slide please' and is well away. (It is elementary to note that this should be followed by the second slide and then by the first slide in rapid succession.)

Another useful technique, best practiced in conjunction with the first, is the 'White Dot Shift.' All slides are of course marked with a white dot in one corner in which the projectionist places his right thumb to ensure the correct orientation of the slide in the slide holder. The present technique is to mark your first slides in an irregular corner thereby ensuring faulty projection. Combined with the foregoing '3-2-1' technique the 'White Dot Shift' makes a devastating beginning. It would, however, count as merely crude (if effective) but for a further development, aimed both at the projectionist and the audience, that it makes possible. This is for the Slidesman to show puzzlement at the continuing confusion and then for light suddenly to dawn: 'Oh! I'm frightfully sorry about those slides being marked in an unusual way,' he calls to the projectionist, 'You see I always take my *own* projectionist with me to *important* conferences' then as an afterthought 'And he's left-handed.' Finally: 'But don't worry, it's only the *first few* that are like that.'

This should immediately be followed by a 'Parity-Non-Conserving Slide' which does not project correctly no matter how placed in the projector. There are many ways of constructing such slides. The simplest and at the same time the subtlest is to letter the slide with letters which are individually the right way round but with the words running from right to left.

Communications direct to the projectionist are always good and should be made in such a manner that it is not immediately clear whether the Slidesman is addressing the projectionist or the audience. Absurd complication in the instructions must be avoided. The Slidesman uses something like: 'After the next slide but two I shall want to look again at the last one but four.' After the next slide: 'I meant of course that slide which was then going to be the last one but four after I had indeed had that which was going to be the next but two, not that which was then the last but four.' Follow this by skipping one slide.

These are elementary techniques but should suffice for most projectionists. Occasionally resistance is offered in which case slightly more advanced methods are available of which I shall mention only two.

The first in the 'Unfocussable Slide.' This consists of two identical glass sheets, each bearing the Figure, that are fixed in precise registration with each other by a small amount of low-melting-point wax. After a short time on the screen the heat of the projection lamp melts the wax and one sheet slips about a millimetre relative to the other thereby throwing the Figure out of vertical focus. The Slidesman's sharp cry of 'Focus, please' rouses the projectionist to frantic, vain and incredulous efforts. The Slidesman's advice of 'No, no, focus it up and down, not side to side' produces a useful effect.

A second advanced technique I have named 'Groshev's Blank Pair' in honour of a great Soviet Slidesman to whose inspiration I owe this important development. It consists of two successive perfectly blank slides. These come at the end of a run of slides that have been taken in very rapid succession, producing an exhausting and hypnotic effect upon the projectionist. Suddenly the run comes to an end and the projectionist, after loading up the next slide as usual, returns with thankfulness to the biting of his nails. However the slide that he projected in response to the last urgent 'New slide please' was in fact the first of the two perfectly blank ones and the second blank slide is now of course also safely loaded up in the slide holder. After several seconds comes the Slidesman's icy 'I said "Next slide please"' and the projectionist sees to his horror that, although he put a slide in and shoved it across, nothing has come onto the screen. He knows he put in a slide; he must then, by some aberration, have taken out the last but one projected slide without replacing it, projected the unfilled holder and replaced instead the last projected slide with the one that should now be showing but in fact is waiting to be shoved across. With a muffled cry and on the third repetition of 'I am still waiting for the next slide' the projectionist slams over the slide holder. The fourth repetition of 'I am still waiting . . .' coincides with the projectionist's first bubbling moan. Although his vision is now clouding he knows how to reassure himself of his sanity. Back again he slams the slide holder, and, his world at stake, thrusts his finger straight at the middle of the second blank slide to verify its tangible existence. Now *there is a large hole in the centre of the second blank slide*. The slide is little more than a rim. The moan swells to full gibber. With the last vestige of his reason the projectionist seizes the next slide in the box and tries to ram it into the holder already occupied by the second blank slide.

So much for the projectionist. I now turn to the more important problem of the audience. This is a far subtler matter. The Slidesman's objective is, of course, to convey, effortlessly, to the members of the audience, his transcendence and superiority over them. This is the full field of Conferencemanship and I must repeat how much I regret that I am unable to introduce all aspects of this study. A guiding principle of Conferencemanship is to conceal from the audience what you are talking about. The Conferenceman as Slidesman must equally conceal what his slides are about. It should, however, be all but clear that whatever the slides are about it is not what the talk is about, whatever that is. The general malaise that this engenders is helped by remarks such as: 'This same point is made diagrammatically in the next slide.'

The only exception to this rule is when the Slidesman is using the 'Time Lapse' technique which is very disturbing. In this the Slidesman says very clearly something extremely simple and extremely lucid as he shows a very simple slide. He then says he wishes to point out the very sharp distinction between that situation and the following one. The next slide is precisely the same as its predecessor and the Slidesman says precisely the same thing as before. This may be repeated several times in succession. It is helped if the Slidesman addresses his remarks specifically towards the most distinguished of those persons in the front row who have just woken up. The distinguished elder will nod in more and more vigorous assent as each fresh distinction is drawn. It may be his last conference.

Slides are very useful for conveying to the audience the Slidesman's togetherness with the Olympians. A good technique is to show a slide with something written with a sticky pencil on the *reverse side* so that the muddy, reversed, letters are made out only with difficulty by the fascinated audience. What the Slidesman has written is: 'Wigner asks for *two* copies of this slide.' The next slide but one bears the inscription (this time it can be on the right side): 'This one for Eugene too.'

The Slidesman-audience relationship is fostered by the 'Interpolated Slide'. This is from a field utterly alien to that of the conference and illustrates, say, a sequence of tablets inscribed in Linear B, or, perhaps, a manuscript page of an unpublished arrangement by Busoni for one piano, three hands, of a motet by Gesualdo. The Slidesman says: 'I'm terribly sorry. It must have crept in somehow' and then after a tiny pause 'Another of my little foibles you know.' The implication that, firstly, this remote subject is but one of an unspecified number of the Slidesman's little foibles (on which he evidently speaks at conferences) and that, secondly, he regards nuclear physics also as a little foible, are both satisfactory.

An arresting technique is the 'Further Work' slide. This shows a number of points labelled 'Experiment' all lying well below a horizontal line labelled 'Theory.' The Slidesman (who is, of course, responsible for both the theory and the experiment) refers to the points as 'Very recent work in my laboratory.' (Always, *my* laboratory) and says that although the present fit between theory and experiment is not of the best further work is going on in his laboratory at that very moment and that he feels confident that when the new results come along the fit between experiment and theory will be greatly improved. As he says this the experimental points, which are really little weights fixed on with more low-melting-point wax, respond to gravity as the wax softens in the heat of the projector and move across the screen to rest finally on the theoretical line which is really a strip sticking out of the surface to make a ledge.

The final section of Slidesmanship on which I shall be able to touch in this all too sketchy survey is an intimate part of that branch of Conferencemanship to do with impressing your audience with the wealth of remote and exotic conferences, of which they have never heard, that you the Conferenceman have attended. This is the heart of Conferencemanship. The Slidesman rises during a discussion and says 'But this matter was absolutely thrashed into the ground at the Addis Ababa Conference.' This is good but it must be pressed home by: 'I happen to have a slide with me that Professor Poop kindly gave me after this meeting. It sums up very nicely and will save us any more discussion.' It does not matter what the slide is about either.

A useful piece of Slidesmanship is to have a run of slides all of which project on their sides with abscissae running vertically. This causes cricked necks in the audience, itself very useful, and enables the Slidesman to say: 'I'm sorry about these slides, they were made up for the Peiping Conference.'*

At one time Russianmanship was an important part of Conferencemanship but now that everybody of any consequence has been to two Russian Conferences this must be largely dropped. Slides actually lettered in Cyrillic characters, however, are still most valuable. If the Slidesman presents a long run of slides, all so lettered, but without translating he implies both that he is so frequent a visitor to Soviet parts that it is worthwhile getting his slides specially made up and also that he is so familiar with the language that it never enters his head that it needs translating. This is good but the best is to come. Eventually someone in the audience must tire of this meaningless procession and say: 'Look here, aren't you going to tell us what those slides are about? We can't *all* read Russian you know.' After a well-judged pause, the Slidesman replies, 'Not *Russian*, my dear fellow, *Bulgarian*.'

* Note to UK readers: Everybody but us and the Chinese use slides that are $3\frac{1}{4}'' \times 4''$; Chinese slides are $4'' \times 3\frac{1}{4}''$.

---

Mathematics are a species of Frenchman; if you say something to them, they translate it into their own language and presto! it is something entirely different.

GOETHE

# A conference glossary

DAVID KRITCHEVSKY and R J VAN DER WAL

From *Proceedings of the Chemical Society* May 1960 p 173.

A. IN PRESENTING PAPERS

| *When They Say:* | *They Mean:* |
|---|---|
| 1. Elegant | A reference to work of an author whose work is to be attacked |
| 2. A surprising finding | We barely had time to revise the abstract. Of course we fired the technician |
| 3. Preliminary experiments have shown that . . . | We did it once but couldn't repeat it |
| 4. The method, in our hands. . . . | Somebody didn't publish all the directions |
| 5. A survey of the earlier literature | I even read through some of last year's journals |
| 6. Careful statistical analysis | After going through a dozen books, we finally found one obscure test that we could apply. |
| 7. We are excited by this finding | It looks publishable |
| 8. We have a tentative explanation | I picked this up in a bull session last night |
| 9. We didn't carry out the long-term study | We like to go home at 5 pm What do you think we are, slaves? |
| 10. The mechanism is not yet clear | We plan to do the second experiment as soon as we get home |

B. IN DISCUSSION

| | |
|---|---|
| 1. We say this with trepidation | (*a*) They are going out on a limb when in the presence of an author whose work is to be, or has been, attacked. |
| | (*b*) They are about to make a statement about something they know nothing about. |

| | |
|---|---|
| 2. Could you discuss your findings? | Tell us now. Don't hide it in some obscure journal |
| 3. Have you considered the possibility? | Have you read my work? |
| 4. Have you any ideas at all . . . | What are you keeping from us? |
| 5. Would you care to speculate? | I wonder if you agree with me |
| 6. Why do you believe . . .? | You're out of your mind |
| 7. I would like to make one comment on these suggestions | Awful! |
| 8. We cannot reconcile these data. | Are you telling the truth? |
| 9. We have repeated your experiments in our lab. | Brother, were we surprised! |
| 10. Did I read your slide correctly? | Did you write it correctly? I never make mistakes |

It is evident that the field of scientific semantics offers ground for fruitful investigation (which means 'I never expect to do it myself, but if someone does, this statement will give me a claim on priority').

## Valentine from a Telegraph Clerk ♂ to a Telegraph Clerk ♀

JAMES CLERK MAXWELL

*'O tell me, when along the line*
  *From my full heart the message flows,*
*What currents are induced in thine?*
  *One click from thee will end my woes'.*

*Through many an Ohm the Weber flew,*
  *And clicked the answer back to me,*
*'I am thy Farad, staunch and true,*
  *Charged to a Volt with love for thee'.*

[*In Maxwell's time, the term Farad was sometimes used for what we now call a Coulomb while a Weber meant an Ampère.*]

# Enrico Fermi

EMILIO SEGRÈ

From *Enrico
Fermi Physicist*
by Emilio Segrè
(University of
Chicago Press
1970) p 134
Oppenheimer had been among the first to introduce quantum
mechanics to America and had founded a flourishing school of
theoretical physics which produced many of the leading American
theoreticians. He often presented physics in rather abstract terms
which contrasted, at least in my mind, with the simple, direct
approach to which Fermi had accustomed me. I remember a
remark that Fermi made in 1940 at the time of his visit to Berkeley
for the Hitchcock lecture. After attending a seminar given by one
of Oppenheimer's pupils on Fermi's beta-ray theory, Fermi met
me and said: 'Emilio, I am getting rusty and old, I cannot follow
the highbrow theory developed by Oppenheimer's pupils any-
more. I went to their seminar and was depressed by my inability
to understand them. Only the last sentence cheered me up; it was:
"And this is Fermi's theory of beta decay."'

---

From *How to tell
the Birds from the
Flowers* (New
York: Dover)
[*R W Wood was well known as an optician, writer of hundreds of papers
notably on resonance radiation. But his most famous book is 'How to tell
the Birds from the Flowers', first published in 1917.*]

## The Parrot.    The Carrot.

*The Parrot and the Carrot one may easily confound,*
*They're very much alike in looks and similar in sound,*
*We recognize the Parrot by his clear articulation,*
*For Carrots are unable to engage in conversation.*

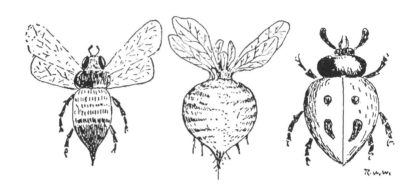

# The Bee. The Beet. The Beetle.

*Good Mr. Darwin once contended*
*That Beetles were from Bees descended,*
*And as my pictures show I think*
*The Beet must be the missing link.*
*The sugar-beet and honey-bee*
*Supply the Beetle's pedigree:*
*The family is now complete,—*
*The Bee, the Beetle and the Beet.*

## Absent-minded

HENRY ROSCOE

From '*Bunsen Memorial Lecture*' *Journal of the Chemical Society* (1900) 77.

Like many men who are engrossed in their special calling, Bunsen was often absent-minded, and many good stories were current about the mistakes which he thus unwittingly made. He had a well known difficulty in remembering names; one day a visitor called who he knew quite well was either Strecker or Kekulé. During the conversation he was endeavouring without success to make up his mind which of these two gentlemen was his caller. First he thought it was Kekulé, then he convinced himself that he was talking to Strecker. At last, however, he decided that it was really Kekulé. So when his visitor rose to take leave, Bunsen, feeling confidence in his last conclusion, could not refrain from remarking, 'Do you know that for a moment I took you for Strecker!' 'So I am,' replied his visitor in amazement.

# The Mason–Dixon line

From *Philosophical Transactions of the Royal Society* (1768).

[*Lands in the American Colonies had been granted to William Penn and Lord Baltimore, known respectively as Pennsylvania and Maryland. But the exact position of the boundary between them (nominally an east–west line at latitude 40° N) was subject to protracted disputes. Eventually in 1763 the proprietors engaged the astronomer Charles Mason and his assistant Jeremiah Dixon to carry out an accurate survey. But this project led to others. One of the urgent scientific issues was the magnitude of the ellipticity of the earth, deduced from measurements of the length of a degree of latitude at different distances from the equator. An introduction to Mason and Dixon's paper explained that the opportunity was taken to survey not only the east–west line but also to survey a north–south line 100 miles long—a project supported by the Royal Society.*]

In the course of this work, Messrs Mason and Dixon traced out and measured some lines lying in and near the meridian, and extended, in all, somewhat more than 100 miles; and, for this purpose, the country in these parts being all over-grown with trees, large openings were cut through the woods, in the direction of the lines, which formed the straightest and most regular, as well as extensive vistas that, perhaps, ever were made.

They perceived that a most inviting opportunity was here given for determining the length of a degree of latitude, from the measure of near a degree and half. And, one remarkable circumstance very much favoured the undertaking, which was, that the country through which the lines run, was, for the most part, as level as if it had been laid out by art.

[*Mason and Dixon's calculations of the north–south distance and of the differences of latitudes ended as follows.*]

The sum is = 538 067 feet = an arch of meridian intercepted between parallels of latitude answering to the celestial arch 1° 28′ 45″. Then say, as 1° 28′ 45″ is to 1° so is 538 067 feet, to 363 763 English feet, which is the length of a degree of latitude in the provinces of Pennsylvania and Maryland. The latitude of the northernmost point was determined from the zenith distances of several stars, = 39° 56′ 19″, and the latitude of the southernmost point = 38° 27′ 34″. Therefore the mean latitude expressed in degrees and minutes is = 39° 12′.

To reduce this measure of a degree to the measure of the Paris toise, it must be premised, that the measure of the French foot was found on a very accurate comparison, made by Mr Graham, of the toise of the Royal Academy of Sciences at Paris, with the Royal Society's brass standard, to be to the English foot, as 114 to 107. Therefore say as 114 is to 107 so is 363 763 the measure of the degree in English feet, to 341427 the measure of the degree in French feet, which divided by 6, the number of feet in a toise, gives the length of the degree = 56904½ Paris toises, in the latitude 39°.

It must however be observed, that the accuracy of this reduction into Paris toises depends on a supposition that the length of the French toise, which is of iron, was laid off by the gentleman of the Royal Academy of Sciences, on the brass rod sent over to them for that purpose by Mr Graham, which was afterwards returned to him, in a room where the heat of the air answered to 62 of Fahrenheit's thermometer, or 15 of Reaumur's, or nearly so, which is probable enough, but is a point that does not appear to have been ascertained. For, on account of the difference of expansion of brass and iron, 2 rods made of those metals, however accurately they may be made of equal lengths at first, will only agree together afterwards in the same temperature of the air in which they were originally adjusted together. It is fortunate that the uncertainty in the present case is but small, since 20° difference of Fahrenheit's thermometer or 10° of Reaumur's, produces, according to Mr Smeaton's experiments, a difference of the expansions of brass and iron, of only 1/13500th part, which would cause an error of only 27 English feet, or about 4 Paris toises in the length of the degree.

It is however to be wished, that the proportion of lengths of the French and English measures might be again ascertained by another careful experiment, in which the temperature of the air, as shown by the thermometer, might be noted at the time.

[*Not only was the original line extended across the country to mark the division between North and South, but the work stimulated accurate measurements of expansion coefficients.*]

Scientists animated by the purpose of proving that they are purposeless are an interesting subject of study.

ALFRED NORTH WHITEHEAD, *The Function of Reason*

One humiliating thing about science is that it is gradually filling our homes with appliances smarter than we are.

ANONYMOUS

ANTIKYTHERA MECHANISM
PARTIALLY RECONSTRUCTED

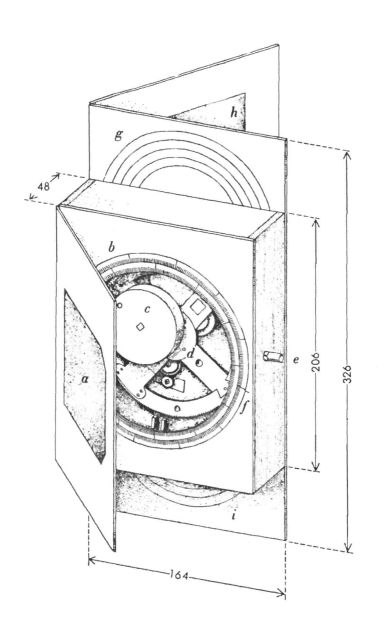

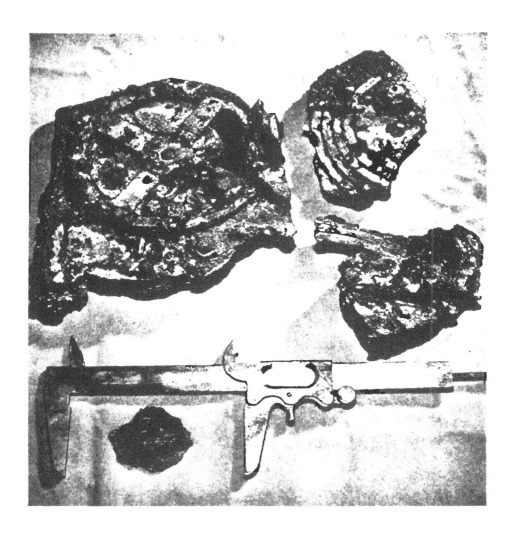

# Toothed wheels

Line illustration from 'An Ancient Greek Computer', Scientific American, June 1959 p 60; Antikythera machine photograph by courtesy of Derek de Solla Price; astrolabe photographs by courtesy of Museum of History of Science, Oxford.

The two instruments shown here are among the most intriguing objects ever studied by historians of science. They could be described as analogue computers for displaying astronomical data—ancestors of the clock. Their fascination lies in the fact that these two are the earliest known devices which incorporated flat, toothed gearwheels. The wheels in both have 60° teeth and square shanks, and they resemble one another so closely that they must be part of the same technological tradition. Yet one is dated 65 (± 10) BC, the other AD 1222. They are separated by thirteen centuries but nothing comparable has yet been found belonging to the interval between.

The Greek computer was discovered in 1901 in the wreck of a treasure ship sunk off the island of Antikythera, between Greece and Crete. The corroded fragments of brass and verdigris have been gradually cleaned and the instructions engraved on some of the plates partly deciphered. The partial reconstruction shows the square-section shaft *e* which was turned, presumably with a key, to rotate the slip rings *g* and *i*, probably indicating the positions of the seven 'planets', and the graduated scale *f* showing the position of the sun in the zodiac.

The astrolabe of Muhammad ibn Abī Bakr the needle-maker of Isfahan in Persia resembles a design described in a manuscript written about AD 1000 by an author who is known to have visited India—perhaps this gives a clue to where the tradition was preserved. The dial has windows showing the year and the phase of the moon, and four slip rings showing planetary positions. Some of the gears are missing.

---

Newspaper report.

Navy volunteers whirl on a merry-go-round, set mark in withstanding gravity forces . . . The capsule was spun on an arm at more than 3000 miles an hour for a period of 12 seconds. ie, at about 1 mi/sec in a 5 mile radius.

---

There is something fascinating about science. One gets such wholesale returns of conjecture out of such a trifling investment of fact.

MARK TWAIN, *Life on the Mississippi*, 1874

173

# The transit of Venus

JEREMIAH HORROX

From *The Gentleman's Magazine* 31, 222 (1761), one of the early translations of the latin *Venus in sole visa,* emended by comparison with A B Whatton's *Memories of the life & labours of Jeremiah Horrox* (London) 1859.

*[One of the first Englishmen to understand Kepler's theory was Jeremiah Horrox, a young curate whose parish was in a low-lying and remote part of Lancashire. He was desperately poor—the living was 'a very poor pittance;' his efforts at teaching were 'daily harrassing duties'. But he bought a telescope and by the age of 20 had established that the moon's orbit is an ellipse with the earth at one focus. Calculating ephemerides, he stumbled across the fact that Venus should pass across the sun's disc on about November 24, 1639 (December 4 on our new style).*

*A transit of Venus had never been observed before. It was important to establish the exact date and time because that would emphasize the inadequacy of geocentric theories and allow Kepler's figures to be refined. Indeed Horrox' data were used extensively by Newton 75 years later.*

*William Crabtree was a linen draper of Manchester and the two were in correspondence.]*

I invited my friend, *Wm Crabtree* of *Manchester* to this *Uranian banquet,* in a letter dated *Hool, Oct.* 26, 1639; who, in mathematical knowledge, is inferior to few. I communicated my discovery to him, and earnestly desired him to make whatever observation he possibly could with his telescope, particularly to measure the diameter of the planet *Venus*; which, according to *Kepler*, would amount to 7 m. according to Lansbergius to 11 m. but which according to my own proportion, I expected would hardly exceed *one* minute. I said, that the conjunction, according to *Kepler*, will be *Nov.* 24, 1639, 8 h. 1 m. A.M. at *Manchester*, the planet's latitude being 14 m. 10 s. south; but according to my own correction, I expected it to happen at 5 h. 57 m. P.M. at *Manchester*, with 10 m. lat. south. But because a small alteration in *Kepler's* numbers would greatly alter the time of the conjunction, and the quantity of the planet's latitude, I advised to watch the whole day, and even on the preceding afternoon, and the morning of the 25th, though I was entirely of the opinion that the transit would happen on the 24th.

After having fully weighed and examined the several methods of observing this uncommon phenomenon, I determined to transmit the Sun's image thro' a telescope into a dark chamber, rather than thro' a naked aperture, a method greatly commended by *Kepler*; for the Sun's image is not given sufficiently large and distinct by the latter, unless at a very great distance from the aperture, which the narrowness of my situation would not allow; nor would *Venus*'s diameter be visible, unless the aperture were very small; whereas, my telescope, which rendered the solar spots distinctly visible, would shew me *Venus*'s diameter well defined, and enable me to divide the Sun's limb more accurately.

Having attentively examined *Venus* with my instrument, I described a circle upon paper, which nearly equalled six inches, the narrowness of the apartment not allowing a larger size; but even this size admitted divisions sufficiently accurate.

When the time of observation drew near, I retired to my apartment; and having closed the windows against the light, I directed my telescope previously adjusted to a focus, thro' the aperture towards the Sun, and received his rays at right angles upon the paper. The Sun's distinct image exactly filled the circle, and I watched carefully and unceasingly for any dark body that might enter upon the disk of light; and tho' I could not expect the planet to enter upon the Sun's disk before three o'clock on the afternoon of the 24th, from my own corrected numbers, upon which I chiefly relied; yet, I observed the Sun on the 23rd, but more particularly on the 24th; for on the 24th I observed the Sun from the time of its rising to 9 o'clock; and again, from a little before ten until noon; and at one in the afternoon, being called in the intervals to business of the highest moment, which for these ornamental pursuits, I could not with decency neglect. But in all these times I saw nothing on the Sun's face except one small and common spot, which I had seen on the preceding day, and which also I afterwards saw on some of the following days.

[*Horrox' biggest difficulty was not so much the cloudy weather but rather that November 24 was a Sunday. He had to take Matins, Holy Communion and Evensong, preaching two massive sermons, during the short day. As time wore on and only 35 minutes were left till sunset, one can guess his emotions.*]

But at 3h. 15m. in the afternoon, when I was again at liberty to continue my labours, the clouds, as if by *Divine Interposition*, were entirely dispersed, and I was once more invited to the grateful task of repeating my observations. I then beheld a most agreeable sight, a *spot*, which had been the object of my most sanguine wishes, of an unusual size, and of a perfectly circular shape, just wholly entered upon the Sun's disk on the left side, so that the limbs of the *Sun* and *Venus* exactly coincided in the very point of contact. I was immediately sensible that this round spot was the planet *Venus*, and applied myself with the utmost care to prosecute my observations.

[*These were the point of entry on to the Sun, the line of motion across it and the diameter of the planet.*]

175

All the observations which could possibly be made in a short time, I was enabled, by *Divine Providence*, to complete so effectually that I could scarcely have wished for a more extended period.

Mr *Crabtree* readily complied with my request and intended to observe the transit in the same manner with me; but the sky was very unfavourable to him, and was so covered with clouds, almost during the whole day, that he gave himself up entirely to despair and resolved to take no further trouble in the matter. But, a little before the time of sun-set, about 3h. 35m. by the clock, the Sun breaking out for the first time from the clouds, he eagerly betook himself to his observation, and happily saw the most agreeable of all sights, *Venus* just entered upon the *Sun*. He was so ravished with this most pleasing contemplation, that he stood for some time viewing it leisurely, as it were; and, from an excess of joy, could scarce prevail upon himself to trust his own senses. For we astronomers have a certain *womanish* disposition, distractedly delighted with light and trifling circumstances, which hardly make the least impression upon the rest of mankind. Which levity of disposition, let those deride that will; and with impunity; and if it gratify them I too will join in the merriment. But let not any severe *Cato* be seriously angry with these vanities of ours: For what youth, such as we are, would not fondly admire *Venus* in conjunction with the *Sun*, what youth would not dwell with rapture upon the fair and beautiful face of a lady, whose charms derive an additional grace from her fortune?—But to return, he from his ecstacy, I from my digression. The clouds deprived Mr *Crabtree* of the sight of the Sun, almost as soon as he was roused from his reverie; so that he was able to observe little more than that *Venus* was certainly in the Sun.

[*But he was able to sketch the positions from memory.*]

I hope to be excused for not informing other of my friends of the expected phenomenon; but most of them care little for trifles of this kind, preferring rather their *hawks* and *hounds*, to say no worse. If others, without being warned by me, have witnessed the *transit*, I shall not envy their good fortune; but rather rejoice, and congratulate them on their diligence.

[*Though Horrox and Crabtree lived only 30 miles apart, both were so poor that they could never undertake such a long journey and they had never met. They planned to do so on January 4, 1641, but Horrox died the day before that, 'very suddenly', aged 22.*]

# Lines inspired by a lecture on extra-terrestrial life

JDGM

From *The Observatory* 65, 88 (1943).

*Some time ago my late Papa*
*Acquired a spiral nebula.*
*He bought it with a guarantee*
*Of content and stability.*
*What was his undisguised chagrin*
*To find his purchase on the spin,*
*Receding from his call or beck*
*At several million miles per sec.,*
*And not, according to his friends,*
*A likely source of dividends.*
*Justly incensed at such a tort*
*He hauled the vendor into court,*
*Taking his stand on Section 3*
*Of Bailey 'Sale of Nebulae.'*
*Contra was cited Volume 4*
*Of Eggleston's 'Galactic Law'*
*That most instructive little tome*
*That lies uncut in every home.*
*'Cease' said the sage 'your quarrel base,*
*Lift up your eyes to Outer Space.*
*See where the nebulae like buns,*
*Encurranted with infant suns,*
*Shimmer in incandescent spray*
*Millions of miles and years away.*
*Think that, provided you will wait,*
*Your nebula is Real Estate,*
*Sure to provide you wealth and bliss*
*Beyond the dreams of avarice.*
*Watch as the rolling aeons pass*
*New worlds emerging from the gas:*
*Watch as the brightness slowly clots*
*To eligible building lots.*
*What matters a depleted purse*
*To owners of a Universe?'*
*My father lost the case and died:*
*I watch my nebula with pride*
*But yearly with decreasing hope*
*I buy a larger telescope.*

# Postprandial: Ions mine

J J E DURACK

[*In the heroic days of the Cavendish Laboratory it was the custom to hold an annual dinner followed by home-made entertainments, usually songs at the piano. These 'Postprandial Proceedings of the Cavendish Society' (Cambridge: Bowes and Bowes 1926) celebrate the discoveries of gas discharge phenomena and the early days of nuclear physics.*

AIR 'Clementine.'

1 *In the dusty lab'ratory,*
  *Mid the coils and wax and twine,*
  *There the atoms in their glory*
  *Ionize and recombine.*

CHORUS *On my darlings! Oh my darlings!*
  *Oh my darling ions mine!*
  *You are lost and gone for ever*
  *When just once you recombine!*

2 *In a tube quite electrodeless,*
  *They discharge around a line,*
  *And the glow they leave behind them*
  *Is quite corking for a time.*

3 *And with quite a small expansion*
  *1 :8 or 1 :9,*
  *You can get a cloud delightful,*
  *Which explains both snow and rain.*

4 *In the weird magnetic circuit*
  *See how lovingly they twine,*
  *As each ion describes a spiral*
  *Round its own magnetic line.*

5 *Ultra-violet radiation*
  *From the arc or glowing lime,*
  *Soon discharges a conductor*
  *If it's charged with minus sign.*

6 α *rays from radium bromide*
  *Cause a zinc-blende screen to shine,*
  *Set it glowing, clearly showing*
  *Scintillations all the time.*

# The trial of Galileo

[*Popular versions of the trial of Galileo represent it as a confrontation of the goodies and the baddies, a champion of scientific truth against a reactionary church. Reality is more complicated. It was in fact university professors who instigated the affair. It was as if in modern terms a proponent of some new version of fundamental particle theory were to have his political allegiance besmirched by his university colleagues. In sixteenth-century Italy the universities were dominated by monks who had a vested interest in protecting the ancient learning because that was inextricably interwoven into religious truth. The personalities of the Church—Pope and Cardinals—were more accommodating to new discoveries; they were brought into the controversy reluctantly, although there is little doubt that eventually they were angered by Galileo's tactlessness and brilliant sarcasm.*

*The following account consists of extracts from the classic* 'Galileo and the Freedom of Thought' *by the chemist F Sherwood Taylor (London: Watts, 1938). In it, the words 'monks' and 'Peripatetics' can be taken to mean the university hierarchy. The controversy was between the old earth-centred system, the finite universe of Ptolemy (AD 140) ostensibly based on the physics of Aristotle, and the sun-centred universe, infinite in extent, proposed by Copernicus in 1543.*

*The controversy was sparked off when in 1609 Galileo made his first telescopes and within a short time observed the mountains of the moon, sunspots, the moons of Jupiter and the phases of Venus. The first two of these threw doubt on the perfection of the heavens, the other two made it plausible that the earth was not a unique planet. In 1613, Galileo moved from a chair of mathematics at Padua in the independent state of Venice, to a similar post at Florence in Tuscany, a less independent state and much nearer Rome.*]

With Galileo's removal to Florence begins the drama of his relations with the Church.

The first serious attack, abortive as it proved, was made in 1610 or 1611 by Ludovico delle Colombe in a pamphlet entitled *Contro il moto della Terra*—Against the Motion of the Earth. Galileo's name nowhere appears in it, but the tenor of the work shows it to be aimed at him and his school.

The author begins by pointing out that the universe cannot be adequately described by mathematics, which is an abstraction from natural phenomena taken as a whole, and may predict many phenomena which are in practice impossible.

He then gives a number of examples of phenomena which he supposes to conflict with the motion of the earth. Thus, he says, suppose a cannon to be fired, first due east, then due west. In one case the shot has the velocity of the earth as well as that of the force of the powder, in the other case only the difference of the two. But, in fact, no such difference is noted, and therefore the earth does not move.

His second argument is similar. If the earth moved, and one were to shoot vertically with a cross-bow so as to make the projectile return to one's feet, it would not return there, whereas the contrary is true.

Thirdly, if the earth rotates so fast as Copernicus says, birds could not keep up with it!

Fourthly, suppose a ball of lead and a ball of cork to be dropped from a height, the former, which would descend faster than the latter, would be left behind by the earth to a less extent, so the two balls would land at different spots, which does not in fact occur.

The idea of the moon being mountainous and composed of the same sort of matter as the earth shocked him deeply. He supposed that the mountains, which Galileo had demonstrated, were only denser portions, and that the whole was a perfect sphere, the apparent irregularities being filled up with a transparent, invisible material!

But the sting of Ludovico delle Colombe's treatise is in its tail. At the end of his argument he quotes a long series of Scriptural texts to show:

(*a*) That the earth does not move.
> 'Who laid down the foundations of the earth, that it should not be removed for ever' (Ps. civ. 5).
> 'The World also shall be stable, that it be not moved' (I. Chron. xvi.30).

(*b*) That the earth is at the centre.
> 'He . . . hangeth the earth upon nothing [that is, at the centre]' (Job xxvi.7).

Here, then, was the case against the Copernican theory. First, a number of flimsy arguments, arising from the Aristotelian ideas of motion; second, a series of Scriptural texts, backed by the interpretation of the Fathers.

Galileo was a trifle alarmed by this attack, for it was no joke to have one's orthodoxy impugned. He wrote to his friend Cardinal Conti to inquire his opinion as to the compatibility of his views with Holy Scripture. The Cardinal replied that the scriptures favoured Galileo's views about the susceptibility of the heavens to change. He regarded a progressive or translational movement of the earth as conformable to scripture, but thought that a daily rotational movement was hard to reconcile with it.

But Galileo was now enjoying a triumph. Towards the end of March, 1611, he went to Rome, and was received with the highest acclamations. Archbishops and Princes of the Church were delighted to witness the new wonders of the sky. Their discoverer received the greatest of scientific honours in being elected to the famous Accademia dei Lincei, which occupied something of the position of the French Academy or the Royal Society.

The new astronomical discoveries were eagerly studied by the highest authorities of the Church, despite the fact that they certainly seemed to support the Copernican theory, and with no uncertain voice showed the ancient conception of the heavens to be untenable. The Peripatetics were naturally much incensed, both by the doctrines and their enthusiastic reception, but they could advance no convincing arguments in favour of the older views.

The position was now exacerbated by a further controversy in which, again, Galileo showed the physics of Aristotle to be erroneous, and at the same time had the opportunity of giving some hard knocks to the same Ludovico delle Colombe who wrote the above-mentioned treatise against the motion of the earth. The controversy occupies some eight hundred large pages of print, and is not of the first interest today. The dispute arose out of the Aristotelian doctrine that ice is formed from water by condensation as a result of cold. But it is well known that ice floats on water. Galileo maintained that anything which floated on water must be lighter than the same bulk of water, and that therefore ice must be less dense than water, and must be formed by rarefaction of, not condensation of water. The reply to this was that bodies do not float only on account of their low 'gravity'—and that, in fact, a flat thin body will float merely in virtue of its shape. Thus a thin slip of ebony can be made to float on water, while a lump of ebony sinks. Surface-tension effects were not well understood, but Galileo showed that this was true only if the surface was unbroken and that such a body would not rise from the bottom of a liquid. Moreover he clearly showed that when a body heavier than water rests on the surface it actually floats well below the surface, so that in fact the floating object may be pictured as a combination of the heavy body and the light air above it but below the general level of the water. Galileo published his conclusions in 1612, and in the same year several Peripatetics, including Ludovico delle Colombe, published replies to it. In 1613 Vincenzio di Grazia also attacked him. Galileo was advised not to reply, but compromised by allowing his favourite disciple, Don Benedetto Castelli, to publish a voluminous counter-blast, in the composition of which, no doubt, Galileo had the chief share.

We must think, then, of this controversy as steadily embittering the relationship between Galileo and the university professors during the years 1611 to 1616. His adversaries were convinced that he was a dangerous man, out to subvert philosophy. We shall not be surprised to find them making use of a weapon which lay close to their hands and to which the letter of the law of their religion entitled—nay, bound them.

It was not long before an attempt was made to impugn Galileo's orthodoxy . . .

[*and there followed three years of intrigues, charges of heresy and other accusations which did not succeed. Galileo talked and wrote too much; he believed implicitly in the power of reason*].

By 1615, the position was that Galileo's doctrines were felt to be dangerous by the reactionary, the timid, and those with a vested interest in Aristotle; that he was practically unassailable by argument, and that he must therefore be silenced by the civil power.

To this end the adversaries of Galileo had shaped a new weapon. On November 13 Father Ferdinand Ximenes denounced Galileo's book on sun-spots to the Inquisitor in Florence. On November 25 it was ordered to be examined, and two propositions were picked out; on February 19 these were sent to the Holy Office for an opinion. On February 24 the qualifiers delivered the following report, the greatest tactical blunder—to say nothing more—that the Inquisition has ever made:

*Propositions to be censured.*

Censure made in the Holy Office of the City,

> Wednesday, February 24, 1616, in presence of the under-signed Theological Fathers.
>
> First: The sun is the centre of the world, and altogether immovable as to local movement.
>
> Censure: All have said that the said proposition is foolish and absurd in philosophy, and formally heretical, in as much as it expressly contradicts the opinions of Holy Scripture in many places according to the proper sense of the words and according to the common explanation and sense of the Holy Fathers and learned theologians.

Second: The earth is not the centre of the world and is
not immovable, but moves as a whole, also with a diurnal
motion.
Censure: All have said that this proposition must receive
condemnation in philosophy; and with respect to theological truth is at least erroneous in faith.

(There follow the signatures of the eleven 'qualifiers'.)

It should be noted that Galileo is not mentioned in this document: it is moreover important to understand its exact significance.

In the first place, the Holy Office allowed itself to make the
error against which Galileo and St Augustine had warned their
readers. It set up as a matter of faith a proposition of natural
science. No more dangerous thing can be done, whether by
Church or dictator, for demonstrable truths will demonstrate
themselves, and in a few years the authority may find itself in a
position from which there is no escape but retreat.

Such an opinion necessarily put a heavy clog on scientific discussion; for years afterwards good Catholics found it wise to steer
clear of astronomical propositions.

On the other hand, Protestant propagandists have much over-
estimated the effect of this opinion. Its effect was, while allowing
the individual to adhere to his own perception of the truth, to
impose on the Catholic the duty of refraining from showing his
dissension from the official view. Its effect was to preclude public
discussion, but not to impose the necessity of interior belief;
somewhat in the same way as a civil servant is restrained from
publishing his politicial views.

From the point of view of the Catholic of the time there was no
enormity in the prohibition of discussion. The truth about natural
things had not yet assumed much importance in the world's eyes;
religious questions, on the contrary, seemed to be of vital import.
The idea of the duty of obedience to the Church had been familiar
for centuries, while the notion of the right to free discussion had
arisen only some eighty years before, and was still largely bound
up with the idea of Protestantism.

The censure having been pronounced, it remained to take practical steps to give it effect. The next day Paul V ordered Cardinal
Bellarmine to summon Galileo to his presence and to admonish
him that he should renounce the condemned opinions, and he
should abstain from teaching or defending this doctrine in any

way whatever; if he were to refuse to agree to this, he was to be imprisoned. Cardinal Bellarmine duly admonished Galileo.

In the Church itself there were many who were opposed to the action taken. Chief among these was Cardinal Barberini, afterwards Urban VIII; and in later years, when he was Pope, the fact that he had disapproved the action of the authorities gave Galileo great hope that the Copernican theory and system might once more receive recognition.

Galileo seems to have been but little awed by this display of the Church's power. It appears from one of his letters that on March 11 he was granted an interview with the Pope, who received him graciously and told him that the calumnies of his enemies would not be lightly believed. This may be the reason why, according to the Florentine Ambassador, Guicciardini, and others, Galileo behaved very rashly in continuing to urge his opinions even after the decree. At any rate he remained in Rome for many weeks, until at the end of May the Grand Duke thought it well to order him to return.

Reading between the lines, we may see in the whole proceedings a victory for the monks, rather unwillingly wrung from the higher officials of the Church. These, therefore, mitigated the blow to Galileo by their favour. He, sanguine as ever, took their approval as a sort of tacit permission to discuss and teach his doctrines as long as he did not publish anything in their defence.

[*But there was an ambiguity. Galileo thought that he had been ordered not to hold or defend the Copernican theory as if it were absolute truth, but that he was allowed to defend it as a hypothesis. In modern terms, he thought he was allowed to propose the sun-centred universe as a possible model, not necessarily the true model. The Inquisition however held that it had ordered Galileo not to teach the theory* in any form whatsoever. *Galileo continued teaching and writing for sixteen years.*

*In 1632, he published the* 'Dialogues concerning the two Principal Systems of the World'; *the arguments for and against the Copernican System are presented as conversations lasting four days, between two interlocutors Salviati and Sagredo and the easily worsted Simplicio. Much of the argument centres on Galileo's theory of the tides, a complicated effect of centrifugal forces, any attraction by the moon being deliberately ruled out.*]

One of the most interesting and hotly debated of problems is the sequence of events which led to the decision to proceed against the Dialogues.

Remembering that Galileo was in high favour with Urban VIII, that the latter had not shown himself averse to novel doctrines, and that the imprimatur had been given to the book; how is it that we find, six months after its appearance, that the Church had taken the decision to prohibit the book and to proceed against the author? It is clear that without the concurrence of Urban VIII no action would have been taken: what was it, then, that led him to this step?

Whether Galileo intended the Dialogues to appear to be a purely hypothetical discussion or not, the world regarded them as a discussion of the real truth of these matters. The book, which was published in February, took some time to reach Rome, and the first two copies arrived there as late as May, to be followed by eighty others in June. Meanwhile the book was being widely read elsewhere; the learned world was divided into partisans of Galileo, who could not find words to express their admiration of the work, and Peripatetics, who found in it a challenge which could not be answered by argument and must be smothered by force, if their system was to survive.

The Dominicans and Jesuits were the driving force of the latter party. The Jesuits had a substantial monopoly of education, and were bound to defend their system to the last: they saw in Galileo's method of argument the prospect of the secularization of science, and they were determined to resist it.

It is clear that the Pope was now convinced that the book contained dangerous doctrine, and that Galileo had used sharp practice in obtaining the imprimatur.

Urban VIII did not place the matter directly in the hands of the Inquisition, but appointed a special commission to consider the book. The reason for this was stated by the Pope himself to be clemency towards Galileo. According to this view, the Pope wished to find some loophole to avoid the necessity of taking action against his old friend. The opposite view, which is at least as probable, is that there seemed to be a difficulty in formulating a sound accusation, and that it was desirable to set out an indictment which would not fail of its object. The members of the commission were selected from Galileo's enemies, and attempts to have two of his friends made members of it were unsuccessful.

The proceedings went on in complete secrecy: Galileo was neither represented nor given any information, so that when, several months later, he appeared before the Inquisition, he had no knowledge of the charges which were to be preferred against him.

After about a month the Commission made its report. It formulated three counts against Galileo himself and eight against the book. The former were:

1. Galileo has transgressed orders in departing from the hypothetical position, in that he has asserted absolutely the mobility of the earth and the fixity of the sun.

2. He has incorrectly deduced the existence of the tides from the non-existent stability of the sun and mobility of the earth.

3. Moreover, he has been fraudulently silent with regard to the command imposed on him by the Holy Office in the year 1616.

The Inquisition summoned Galileo to Rome. He made every effort and used all the influence at his command to avoid or delay the journey. Seventy years of age, in poor health, and only recently recovered from a painful inflammation of the eyes, small wonder if he dreaded the long journey, made worse by the irksome quarantine regulations of a plague-stricken country—a journey, moreover, to a tribunal whose ruthlessness and power he knew too well. . . .

[*There followed four trials in all, dragged out over several months though Galileo was allowed to stay comfortably housed as they proceeded. Galileo himself became convinced that the Inquisition had in fact in 1616 ordered him not to teach Copernican astronomy even as a hypothesis, and he had obviously disobeyed. Eventually he prepared to recant his views, abjectly but probably insincerely*].

To those who wish to set up Galileo as a plaster saint of science, this disavowal of his life's work is inexpressibly shocking. We must not forget that he was old and ill; that he was a good Catholic, and saw in the Church the fountain-head of all that was good. He was not in the position of the martyr, whose burning integrity no power can break. The conflict between the cause of scientific truth and the right of the Church to dictate its members' beliefs raged in his heart, no less than in the world outside.

Now, after three centuries of science, the idea of scientific truth is a clear-cut conception: but, in 1633, Galileo was one of the very few who had separated natural truths from spiritual; it is doubtful if at that time there was a man who would have laid down his life for a scientific hypothesis.

Galileo was forced to kneel and recant his opinions in the following terms:

'I, Galileo Galilei, son of the late Vincenzio Galilei, Florentine, aged seventy years, arraigned personally before this tribunal . . . I adjure, curse and detest the errors and heresies and I swear that in future I will never again say or assert, verbally or in writing, anything that might furnish occasion for a similar suspicion regarding me. . . . I, Galileo Galilei, have abjured as above with my own hands'.

There is a famous story that Galileo, on rising from his knees, muttered *Eppur si muove*—'None the less, it moves!'. This story was accepted in the eighteenth and early nineteenth centuries, but the more critical biographers very naturally rejected it, for the earliest mention of it appeared to be as late as 1778. But recently Fahie, in his *Memorials of Galileo Galilei*, has shown that these words are inscribed on a portrait, which appears to have been painted in the year of Galileo's death. The improbability of his having pronounced these words audibly is extreme, for having submitted with humiliating completeness, it is unlikely that in the same breath he should defy his all-powerful judges. May we conjecture that he told some favourite disciple that he had murmured the phrase inaudibly; and that the story, like a thousand others, became more striking the more often it was told?

## Newton and Facts

From D Bentley, *Memoirs of Sir Isaac Newton 2*, p 407.

John Conduitt, a personal friend of Newton, tells the following: Mr Molyneux related to us that after he and Mr Graham and Dr Bradley had put up a perpendicular telescope at Kew, to find out the parallax of the fixed stars, they found a certain nutation of the Earth which they could not account for, and which Molyneux told me he thought destroyed entirely the Newtonian system; and therefore he was under the greatest difficulty how to break it to Sir Isaac. And when he did break it by degrees, in the softest manner, all Sir Isaac said in answer was, when he had told him his opinion, 'It may be so, there is no arguing against facts and experiments', so cold was he to all sense of fame at a time when a man has formed his last understanding.

# John Dalton's discovery of his colour blindness

From *Memoirs of the Manchester Literary and Philosophical Society* 5, 28 (1798).

[*In French, colour blindness is called* le Daltonisme. *This account of the discovery of his condition was published in 1798.*]

I was always of opinion, though I might not often mention it, that several colours were injudiciously named. The term *pink*, in reference to the flower of that name, seemed proper enough; but when the term *red* was substituted for pink, I thought it highly improper; it should have been *blue*, in my apprehension, as pink and blue appear to me very nearly allied; whilst pink and red have scarcely any relation.

In the course of my application to the sciences, that of optics necessarily claimed attention; and I became pretty well acquainted with the theory of light and colours before I was apprized of any peculiarity in my vision. I had not, however, attended much to the practical discrimination of colours, owing, in some degree, to what I conceived to be a perplexity in their nomenclature. Since the year 1790, the occasional study of botany obliged me to attend more to colours than before. With respect to colours that were *white, yellow,* or *green*, I readily assented to the appropriate term. *Blue, purple, pink,* and *crimson* appeared rather less distinguishable; being according to my idea, all referable to *blue*. I have often seriously asked a person whether a flower was blue or pink, but was generally considered to be in jest. Notwithstanding this, I was never convinced of a peculiarity in my vision, till I accidentally observed the colour of the flower of the *Geranium zonale* by candle-light, in the autumn of 1792. The flower was pink, but it appeared to me almost an exact sky-blue by day; in candle-light, however, it was astonishingly changed, not having then any blue in it, but being what I called red, a colour which forms a striking contrast to blue. Not then doubting but that the change of colour would be equal to all, I requested some of my friends to observe the phenomenon; when I was surprised to find they all agreed, that the colour was not materially different from what it was by day-light, except my brother, who saw it in the same light as myself. This observation clearly proved, that my vision was not like that of other persons.

# Paris, May 1832

IAN STEWART

From *Galois Theory* (London: Chapman and Hall) 1972.

[*Evariste Galois (born 1811 near Paris) was one of the most original pure mathematicians of the early nineteenth century. By the age of 18 he was submitting papers to the Academy of Sciences but they were rejected or lost. In the riots of 1830 Galois was expelled from the Ecole Normale for writing a blistering attack on the director of the school who had prevented the students from taking part in the revolt.*]

On 17th January 1831 he sent once more a memoir to the Academy: *On the conditions of solubility of equations by radicals.* Cauchy was no longer in Paris, and Poisson and Lacroix were appointed referees. After two months Galois had heard no word from them, and he wrote to the President of the Academy asking what was happening. He received no reply.

He joined the artillery of the National Guard, a Republican organization. Soon afterwards its officers were arrested as conspirators, but acquitted by the jury. The artillery was disbanded by royal order. On 9th May a banquet was held in protest; the proceedings became more and more riotous, and Galois proposed a toast to Louis-Philippe with an open knife in his hand. His companions interpreted this as a threat on the king's life, applauded mightily, and ended up dancing and shouting in the street. The following day Galois was arrested. At the trial he admitted everything, but claimed that the toast proposed was actually to 'To Louis-Philippe, *if he turns traitor*', and that the uproar had drowned the last phrase. The jury acquitted him, and he was freed on June 15th.

On 4th July he heard the fate of his memoir. Poisson declared it 'incomprehensible'. The report ended as follows.

'*We have made every effort to understand Galois' proof.* His reasoning is not sufficiently clear, sufficiently developed, for us to judge its correctness, and we can give no idea of it in this report. The author announces that the proposition which is the special object of this memoir is part of a general theory susceptible of many applications. Perhaps it will transpire that the different parts of a theory are mutually clarifying, are easier to grasp together rather than in isolation. We would then suggest that the author should publish the whole of his work in order to form a definitive opinion. But in the state which the part he has submitted to the Academy now is, we cannot propose to give it approval'.

On July 14th Galois was at the head of a Republican demonstration, wearing the uniform of the disbanded artillery, carrying a knife and a gun. He was arrested on the Pont-Neuf, convicted of illegally wearing a uniform, and sentenced to six months' imprisonment in the jail of Sainte-Pélagie. He worked for a while on his mathematics, then in the cholera epidemic of 1832 he was transferred to a hospital. Soon he was put on parole.

Along with his freedom he experienced his first and only love affair, with one Mlle Stephanie D. The surname is unknown; it appears in one of Galois' manuscripts, but heavily obliterated. There is much mystery surrounding this interlude, which has a crucial bearing on subsequent events. Fragments of letters indicate that Galois was rejected and took it very badly. Not long afterwards he was challenged to a duel, ostensibly because of his relationship with the girl. Again the circumstances are veiled in mystery. One school of thought asserts that the girl was used as an excuse to eliminate a political opponent on a trumped-up 'affair of honour'. In support of this we have the express statement of Alexandre Dumas (in his *Mémoires*) that one of the opponents was Pechéux D'Herbinville. But Dalmas cites evidence from the police report suggesting that the other duellist was a Republican, possibly a revolutionary comrade of Galois', and that the duel was exactly what it appeared to be. And this theory is largely borne out by Galois' own words on the matter:

'I beg patriots and my friends not to reproach me for dying otherwise than for my country. I die the victim of an infamous coquette. It is in a miserable brawl that my life is extinguished. Oh! why die for so trivial a thing, for something so despicable! . . . Pardon for those who have killed me, they are of good faith'.

On the same day, May 29th, the eve of the duel, he wrote his famous letter to his friend Auguste Chevalier, outlining his discoveries, later published by Chevalier in the *Revue Encyclopédique*. In it he sketched the connection between groups and polynomial equations, stating that an equation is soluble by radicals provided its group is soluble. But he also mentioned many other ideas, about elliptic functions and the integration of algebraic functions; and other things too cryptic to be identifiable. It is in many ways a pathetic document, with scrawled comments in the margins: 'I have no time!'

The duel was with pistols at 25 paces. Galois was hit in the stomach, and died a day later on 31st May of peritonitis. He refused the office of a priest. On 2nd June 1832 he was buried in the common ditch at the cemetery of Montparnasse.

His letter to Chevalier ended with these words:

'Ask Jacobi or Gauss publicly to give their opinion, not as to the truth, but as to the importance of these theorems. Later there will be, I hope, some people who will find it to their advantage to de-cipher all this mess. . . .'

[*Three days after Galois' funeral, a major riot broke out in Paris. Captain Sadi Carnot, at 36 retired from the Army to devote himself to physics, went to watch the disturbances.*]

From *Reflexions sur la Puissance Motrice du Feu* ed. Hippolyte Carnot (1878).

An officer leading a cavalry charge was galloping along the street brandishing his sabre and knocking people down. Sadi threw him-self at him dodging easily under his arm, grabbed him by the leg the toppled him to the ground. He laid him in the gutter and went on down the street, avoiding the cheers of the crowd.

[*Two months later Sadi was dead of cholera.*]

---

## Pulsars in poetry

JAY M PASACHOFF

From *Physics Today* 22, 19 (1969).

*Twinkle, twinkle, pulsing star*
*Newest puzzle from afar.*
*Beeping on and on you sing—*
*Are you saying anything?*
*Twinkle, twinkle more, pulsar,*
*How I wonder what you are.*

---

## Clouds, 1900

LORD KELVIN

Slightly con-densed from *Philosophical Magazine* (6) 2, 1 (1901).

The beauty and clearness of the dynamical theory, which asserts heat and light to be modes of motion, is at present obscured by two clouds.
I. The first involves the question, How could the earth move through an elastic solid, such as essentially is the luminiferous ether?
II. The second is the Maxwell–Boltzmann doctrine regarding the partition of energy.

[*Kelvin could certainly recognize the important clouds. One needed relativity, the other quantum theory, to blow it away.*]

# An awkward incident

SIR W L BRAGG

I had been making a number of films, demonstrating experiments, for distribution to schools. In June the producer said to me 'That shot of so-and-so which we made three months ago has turned out to be faulty. Will you please come on Monday wearing the same grey suit with a pin-stripe, and we'll repeat that bit'. I had to say 'My wife declared she could no longer stand seeing me in that suit and has given it to a jumble sale'. The producer said 'You *must* get it back. The only alternative is to re-do the whole sequence and that would be prohibitively expensive.' So, on going to our country place, the site of the jumble sale, we asked our cleaning lady—who always knew everything—whether she could cast any light on the disposal of the suit. She said she thought she had seen Mrs S— of N— (a neighbouring village) going off from the sale with a grey suit. With some difficulty we traced Mrs S- down, 'Yes, her husband had bought a grey coat and vest, but he had not bought the trousers; another gentleman had bought the trousers because her husband had not thought he would have any use for them.' During this conversation I caught a glimpse of Mr S— in a neighbouring room having his tea, and it was obvious from his figure that my trousers would not have been useful. Pressed, she was uncertain who the other gentlemen was. So I borrowed the 'coat and vest' and told the producer that he must take the missing shot from the waist up. He said that this difficulty was unique in his experience.

## Shoulders of giants

Newton: letter dated 5 Feb 1676 quoted in *On the shoulders of giants* by Robert K Merton (The Free Press) 1965.

In a letter to Robert Hooke, Newton wrote: 'If I have seen a little further it is by standing on the shoulders of Giants'.

As Gerald Holton, Derek de Solla Price and others have estimated, some 80 to 90 per cent of all scientists known to history are alive today. To capture the very modern thought that much of what is now known in science has been discovered in our own time, Holton introduced a symposium at which distinguished physicists were to report the history of their major discoveries with the disastrously mixed metaphor: 'In the sciences, we are now uniquely privileged to sit side by side with the giants on whose shoulders we stand'.

# Rotating dog

From 'Reminiscences of William Garnett', Nature 128 605 (1931). It was during the visit of the comet in 1874, when unfortunately the comet's tail was a subject of general conversation, that Maxwell's terrier developed a great fondness for running after his own tail, and though anyone could start him, no one but Maxwell could stop him until he was weary. Maxwell's method of dealing with the case was, by a movement of the hand, to induce the dog to revolve in the opposite direction and after a few turns to reverse him again, and to continue these reversals, reducing the number of revolutions for each, until like a balance wheel on a hair spring with the maintaining power withdrawn, by slow decaying oscillations the body came to rest.

## Answer man

From The Answer Man WOR, New York 18, NY. QUESTION: Explain why an ice boat may sail faster than the speed of the prevailing wind.

ANSWER: Dear Friend: In the first place there is very little resistance to an ice boat before the wind. But a sailboat in water has to push a great deal of water aside in order to proceed. Another factor which allows a higher terminal velocity for an ice boat is momentum. You see, as the wind blows the speed of the boat multiplies until the weight of the boat plus the speed of the wind add up to the total speed of the wind.

Actually the percentage of speed developed by an ice boat is not a great deal more than the propelling wind on account of the pull of gravity. Perhaps 15 per cent greater speed would be considered exceptional.

Write again.

## Home run

Newspaper cutting. Correspondent says the ball has not changed . . . All right, so maybe the 1950 baseball isn't any livelier than the 1940 or 1930 baseball. Then what accounts for the outbreak of home runs, of triples and doubles?

To solve the mystery, how about considering the bat? It is being used differently—and more effectively. And it has gradually become smaller.

Here is the amazing thing. You might think that the bigger the bat the more distance a man could get with hitting a baseball. Such is not the case. The distance, or force, that is given to an object is covered by a little formula out of high school physics. This says: Kinetic energy equals one half the mass times the velocity squared.

Note that 'squared'. That's the key word. For example, if the mass in this formula was expressed by the figure 5, and the velocity by the same figure, then if you increased the mass to six you would have only a slight increase in the distance a ball would go. But if you increased the velocity to six, you would have a very considerable increase. Because five squared is 25, while 6 squared is 36. It is as simple as that.

Or to put it another way, you can get a lot more distance on a hit by swinging a small bat very fast than by swinging a big bat somewhat slower. That is why before a game a fungo bat is used to knock balls out to the outfielders for practice. This is a very slender bat. Armed with it, even the weakest hitter can belt one 300 feet with ease. (It could not be used in a game because it would break under the force of a pitch.)

From *Physicists continue to laugh*, MIR Publishing House, Moscow 1968. Translated from the Russian by Mrs Lorraine T Kapitanoff.

I have finished my course work, Professor.

Newspaper report  The more prosaic explanation of scientists for the copper appearance of the moon during total eclipse is that the sunlight which filters through the earth's atmosphere and reaches the moon at this time contains a more reddish tint than ordinary sunlight because the red element of the light predominates.

# The pulsar's Pindar

DIETRICK E THOMSEN and JONATHAN EBERHART

From *Science News* 93 (15 June 1968) p 562.

[*This poem has been dedicated to S Jocelyn Bell of the Cambridge University Mullard Radio Astronomy Observatory, 'whose persistence led astronomy's most awesome personages to the pulsar's puzzling performance. It was written by Dietrick E Thomsen and Jonathan Eberhart in 1968, in the early days of pulsar investigation, when the nature of these heavenly bodies was very mysterious.*]

*Rhythmically pulsating radio source,*
*Can you not tell us what terrible force*
*Renders your density all so immense*
*To account for your signal so sharp and intense?*

*Are you so dense that no matter you own;*
*Not atoms nor protons, save neutrons alone?*
*And do you then fluctuate once every second*
*So fixed that by you all our clocks might be reckoned?*

*Or are you two stars bound together in action*
*That spin like a lighthouse beam gone to distraction?*
*What in the world can account for your course*
*O rhythmically pulsating radio source?*

*And perhaps is there more than your radio beam?*
*Perhaps visible light in a radiant stream?*
*And what if the cause of your well-metered twitch*
*Is a strange but intelligent hand at the switch?*

*A world of astronomers ponder, a-pacing,*
*The cause of your infernal, rock-steady spacing,*
*To see your pulse vary, they valiantly strive,*
*From 1·3372795.*

*But the biggest of mysteries plaguing our earth*
*Is, how of your kind can there be such a dearth?*
*In infinite space one should find ever more;*
*Can it be that your number indeed is but four?*

---

STREPSIADES: But why do they look so fixedly on the ground?
DISCIPLE OF SOCRATES: They are seeking for what is below the ground.
STREPSIADES: And what is their rump looking at in the heavens?
DISCIPLE: It is studying astronomy on its own account.

Aristophanes, *The Clouds.*

# Walter Nernst

EDGAR W KUTZSCHER

Lecturing on the fundamentals of radio wave transmission, Nernst told the story that he had the honour to demonstrate radio transmission to the German Emperor and Empress. The transmitter was in the Institute of Physics and Nernst was supervising the transmission where they selected a phonograph record with a song by the famous Italian tenor, Enrico Caruso. After the transmission, Nernst was asked to come to the Castle. The Empress congratulated him on the wonderful demonstration and said, 'By the way, Good Professor, we did not know that you are such a fine singer!'

From *Electronic Age* 28, No 3 (1969).

"Don't just sit there! If you've processed all the data there is, go out and find more data!"

# Self-frustration

R V JONES

From 'Impotence
and Achievement
in Physics and
Technology',
Nature 207 120–5
(1965).

Self-frustration can arise very easily in military security. Lord
Cherwell, for example, during the War held a pass that was intended to allow him to enter any Government establishment. Such
a universal and vital pass could only be issued to persons of
extreme responsibility; most people held passes which would admit them to only one specific establishment. Since there were few
persons of comparable responsibility to Lord Cherwell, there were
very few of his kind of pass; and its appearance was kept confi-
dential, so that it could not easily be forged. It followed that very
few guards at the entrances to establishments had ever seen it, and
so Lord Cherwell had great difficulty in persuading them that it
was genuine, and to let him in.

There is often peculiar humour about self-frustration. Consider,
for example, a train of events which started outside the old Clarendon Laboratory, Oxford. I came across a dirty beaker full of
water just when I happened to have a pistol in my hand. Almost
without thinking I fired, and was surprised at the spectacular way
in which the beaker disappeared. I had, of course, fired at beakers
before; but they had merely broken, and not shattered into small
fragments. Following Rutherford's precept I repeated the experiment and obtained the same result: it was the presence of the
water which caused the difference in behaviour. Years later, after
the War, I found myself having to lecture to a large elementary
class at Aberdeen, teaching hydrostatics *ab initio*. Right at the
beginning came the definitions—a gas having little resistance to
change of volume but a liquid having great resistance. I thought
that I would drive the definitions home by repeating for the class
my experiments with the pistol, for one can look at them from the
point of view of the beaker, thus suddenly challenged to accommodate not only the liquid that it held before the bullet entered it,
but also the bullet. It cannot accommodate the extra volume with
the speed demanded, and so it shatters.

The experiment became duly public in Aberdeen, and inspired
the local Territorial contingent of the Royal Engineers, who used
sometimes to parade on Sundays to practise demolition. One task
that fell to them or, more accurately, refused to fall to them, was
the demolition of a tall chimney at a local paper works. There are
various standard procedures for this exercise, one of the oldest
being to remove some of the bricks of one side, and to replace
them by wooden struts. This process is carried so far as to remove
the bricks from rather more than half-way round the base of the
chimney and to a height comparable with the radius. A fire is then

lit in the chimney, to burn through the struts and cause the chimney to fall.

The Royal Engineers, however, decided this time to exploit the incompressibility of water as demonstrated by my experiment. Their plan was to stop up the bottom of the chimney, fill it with water to a height of 6 ft or so, and simulate the bullet by firing an explosive charge under the water. Since diversions on the Sabbath were rare in Aberdeen, the exercise collected a large audience and the charge was duly fired. It succeeded so well that it failed completely. What happened was that, as with the beaker, every brick in contact with the water flew outwards, leaving a slightly shortened chimney with a beautifully level-trimmed bottom 6 ft up in the air. The whole structure then dropped nicely into the old foundation, remaining upright and intact—and presenting the Sappers with an exquisite problem.

Here again, in technology as well as in administration 'the best laid schemes . . . gang aft a-gley' through some element built into the original plan. As a passing example, we may note the failure of all attempts to make a golf ball attain maximum range by polishing its surface. There is a story that P G Tait, then professor of natural philosophy in the University of Edinburgh, calculated the maximum range of a golf ball, and that his son then took him out on the links and showed him that the ball could be driven much further. It was also said that old balls went further than new ones. This was probably true, because the surfaces of old balls were chipped and roughened; and, based on this observation, new balls have ever since been intentionally dimpled. The then unknown factor which later came to light in the wind tunnel was that the roughness encouraged the onset of small-scale turbulence around the ball, and—over the useful range of velocities—this forestalled and obviated the large-scale and more dissipative turbulence which would occur when the laminar flow around the smooth ball ultimately broke down. Rough balls thus have less resistance, paradoxically, than smooth ones. There is probably a lesson for the administrators here as telling as anything in classical politics.

[*Archaeologists sometimes find that the Romans carefully puckered the surface of their sling shots, presumably having found that they travel further.*]

# Unsung heroes—I: J-B Moiré

SIMPLICIUS

From *NPL News* 207 (21 July 1967) pp 10·11.

Moiré fringes are known to everybody, and I have many times been asked who gave his name to them. As a result, I have done a little historical research, and I am now able to give a few fragmentary details of the life of this remarkable inventor.

Jean-Baptiste Moiré was born in 1835 to Aristide Moiré and his wife Therese, née Dubonnet. Therese brought great wealth to the family, deriving from the sale of the remarkable liquid which was invented by her father and bears his name. Their easy style of life enabled Jean-Baptiste to develop to the full his remarkable mental powers.

Moiré was a true polymath, at home alike in the arts and the sciences. He might, in fact, have been termed the Leonardo of the Second Empire had not Leonardo already been christened the Moiré of the Renaissance by the intelligentsia of the Left Bank.

His attention was first drawn to the interaction between gratings by a chance observation at the age of eight, while hanging upside down from the back of the family coach (a frequent practice of his, both in childhood and in later life). The spokes of the turning wheels of the coach, while passing a picket fence, gave rise to striking patterns which made an indelible impression on the mind of the young savant, which impression was reinforced when, in his excitement, he relaxed the grip of his toes and was projected head first onto the adamantine flints of the road.

These patterns remained at the back of his mind throughout adult life, in spite of such multifarious activities as predicting a trans-Neptunian planet (with an orbit at right angles to the ecliptic) and leading an expedition to the South Pole (by correspondence). Ultimately, he wrote a book on these matters, '*Sur les franges des reseaux croissés*,' which was never published. The reason for this lack of publication was that this restless genius became engaged in the problem of the Egyptian hieroglyphs, unaware that it had already been solved by a fellow countryman, Champollion. He effected the translation of the Rosetta Stone with contemptuous ease and, as an academic exercise, rewrote his book in hieroglyphs.

His activities came to the notice of the Society of Antiquaries in London, who invited him over to deliver a paper on this matter. While walking down Piccadilly he encountered his old friend Professor Eddy of the University of Bletchley, the discoverer of the Eddy currents well known to generations of electrical engineers the world over. Engaged in animated conversation, they entered the forecourt of Burlington House; Moiré made to the

199

left in the direction of the rooms of the Society of Antiquaries, while Eddy attempted to turn to the right to the Royal Society.

Predictably, these opposing tendencies cancelled, with the result that the two philosophers walked straight ahead into the Royal Academy, where the Academicians were assembled to hear a lecture on the Pre-Raphaelites. Ever direct in his actions, Moiré strode to the rostrum and began his own lecture.

Unhappily, he commenced by exhibiting an enlarged page of his book, which also contained a picture of the fringes given by two circular gratings. Faced by what seemed to be a new and radical tendency in art, a howl of execration arose from the audience, who rose and drove Moiré from the building.

Pursued by the flower of the artistic Establishment, he fled down Piccadilly. His display card flew from his nerveless fingers and was torn to pieces by the enraged Academicians: however, the fragments were gathered up by a passing Japanese silkweaver, one Hideo Nakamura, who saw the commercial possibilities of these patterns and, home in Japan, launched the now famous moiré silk. Unfortunately, his taste was so bad that his products gave rise to the adjective 'hideous'.

Moiré, always resourceful, made for the Ritz, where his life-long friend, Georges Canapé (Moiré was once a suitor for the hand of his daughter Omelette) was head chef. He took refuge there till, after an urgent message to the Prime Minister, a destroyer took the distinguished savant across the Channel. Unhappily, his scientific curiosity in the action of the rudder prevented him from completing the journey, and so this shining chapter in the history of natural philosophy was prematurely closed.

Each physicist, with a girl beside him, spends two hours a day scanning photographs and gets through 400 or 500

*'The Hunting of the Quark', Sunday Times,* 1 March 1964

From a student essay

It was, in fact, the investigation of heat conduction in a taurus which led Fourier to the discovery of his celebrated series.

# Unsung heroes—II: Juan Hernandez Torsión Herrera

COL. DOUGLAS LINDSAY and CAPT. JAMES KETCHUM

From *Journal of Irreproducible Results* 10 43 (1962).

It is regrettable that there are many men of science who today are almost forgotten. There is indeed only a pitiful handful of scientists and engineers who can quote more than a scrap of biography on Placide Torque, inventor of the Torque dynamometer; or Cholmondeley Bartholomew String, who developed the String galvanometer; or even the contemporary industrial designer Ole Bjerkan, the man to whom we owe the indespensable Bjerkan opener. But there are others too.

Of Juan Hernandez Torsión Herrera very little is known. He was born of noble parents in Andalusia about 1454. He travelled widely and on one of his journeys in Granada with his cousin Juan Fernendez Herrera Torsión both were captured by Moorish bandits. Herrera Torsión died in captivity but Torsión Herrera managed to escape after a series of magnificent exploits of which he spoke quite freely in his later years. During these years he was affectionately known as the 'Great Juan' or as the 'Juan who Got Away'.

Although not a scientist in his own right, Torsión Herrera passed on to a Jesuit physicist the conception of his famous Torsión balance. The idea apparently came to him when he observed certain deformations in the machinery involved when another cousin, Juan Herrera Fernandez Torsión was being broken on the rack.

---

## Wolfgang Pauli

EUGENE P WIGNER

Pauli was a brilliant lecturer if he prepared his address. Once, when I invited him to address our colloquium in Princeton, he did not. The audience became restless and, feeling somewhat responsible for the event, I wanted to help out. He did not define the mathematical symbols he used and I thought that if he explained them, it would help us to understand what he was trying to present.

'Pauli,' I said, 'could you tell us again what your small $a$ stands for?' (The 'again' was sheer politeness; he had not in fact defined it.)

Pauli was flabbergasted by my question and stood there speechless for a few seconds. However, he recovered.

'Wigner,' he said, 'you just have to know everything.' The audience did not laugh.

# Scientific method

ADOLPH BAKER

From *Modern Physics and Antiphysics* (Addison-Wesley) 1970 p 6. The nature of scientific method is such that one must suppress one's hopes and wishes, and at some stages even one's intuition. In fact the distrust of self takes the form of setting traps to expose one's own fallacies. Only when a successful solution has been found can one be permitted the luxury of deciding whether the result is pleasant or useful. The student of physics has his intuition violated so repeatedly that he comes to accept it as a routine experience. When quantum mechanics was first developed in the 1920's in order to explain what had been observed in the laboratory, the implications were extremely painful to the physicists. What had come to be (and are still thought by most people to be) basic principles of scientific philosophy had to be reluctantly abandoned.

Most laymen, when they contemplate the effect physics may have had upon their lives, think of technology, war, automation. What they usually do not consider is the effect of science upon their way of reasoning. Psychiatrists interpret much of the instability in the world today as a product of the destruction of man's myths, which have always been a source of security. Among these were the myth of absolute truth and absolute right, the myth of determinism and predictability, and particularly the myth of the infallibility of established authority, including finally the authority of science itself.

It is customary to blame our sociological problems on the technological fallout resulting from the scientific revolution, but the really villainous act of science was the destruction of these myths. Furthermore, this time there are not even any new myths to replace the old ones. Man has recently discovered that the universe is not the beautifully structured machine his father and grandfather thought they lived in, and he is still reeling from the blow.

# Sir Isaac Newton, a short time before his death

'I do not know what I may appear to the world, but to myself I seem to have been only like a boy playing on the sea-shore, and diverting myself in now and then finding a smoother pebble or a prettier shell than ordinary, whilst the great ocean of truth lay all undiscovered before me.'

# Acknowledgments

The Institute of Physics and the compilers gratefully acknowledge permission to reproduce copyright material listed below. Every effort has been made to trace copyright ownership and to give accurate and complete credit to copyright owners but if, inadvertently, any mistake or omission has occurred, full apologies are herewith tendered.

Addison-Wesley Publishing Co: from *Modern Physics and Antiphysics* by Adolph Baker

American Association for the Advancement of Science: from *Science*

American Institute of Physics: from *Review of Scientific Instruments, Journal of the Optical Society of America, American Journal of Physics, Physics Today*

American Society for Metals: from *Metal Progress*

Pamela Anderton: *Which units of length?*

Associated Book Publishers, Chapman and Hall Ltd and the author: from *Galois Theory* by Ian Stewart

Barnes and Noble Inc and Blackwell and Mott Ltd: from *Flatland, a romance of many dimensions* by Edwin A Abbott

Blackwell and Mott Ltd: illustration from *The Prose Works of Jonathan Swift*

Bowes and Bowes Ltd: from *Postprandial Proceedings of the Cavendish Society*

California Alumni Association and John H Lawrence: from *California Monthly*

Cambridge University Press and the author: from *One Story of Radar* by A P Rowe

H B G Casimir: speech

The Chemical Society: from their *Proceedings*

The University of Chicago Press: from *The Astrophysical Journal* (copyright 1945) and *Enrico Fermi, Physicist* (copyright 1970)

Joel E Cohen: from *On the nature of mathematical proofs*

Columbia University Press: from *University Records of the Middle Ages* by Lyn Thorndike

Dover Publications Inc: from *How to tell the birds from the flowers* by R W Wood

P J Duke: *Snakes and Ladders*

V E Eaton: *Yes, Virginia*

European Organization for Nuclear Research: from *CERN Courier*

Laura Fermi: *Arrogance in physics*

Gauthier Villars: from *La Conférence Solvay (Paris 1921)*

Reinhold Gerharz: from *Solar eclipse*

Donald A Glaser: *Getting bubble chambers accepted by the world of professional physicists*

Harcourt Brace Jovanovich Inc: from *Dr Wood* by William Seabrook, (copyright 1941 by the author, 1969 by Constance Seabrook)

Her Majesty's Stationery Office: from *Admiralty Handbook of Wireless Telegraphy* (*1931*)

Institut de Physique Théorique, Lausanne: from *Helvetica Physica Acta*

The Institute of Electrical and Electronics Engineers Inc: from *Proceedings of the IRE*

The Institution of Electrical Engineers: from *Students' Quarterly Journal*

Malcolm Johnson: *Textbook selection*

R V Jones: *The theory of practical joking, Building research* and *Self-frustration*

Paul Kirkpatrick: *H A Rowland*

Paul E Klopsteg and the publishers: from *Potpourri and Gallimaufry* (copyright 1963 by the American Association for the Advancement of Science)

David Kritchevsky and R J Van der Wal: *A conference glossary*

Edgar W Kutzscher: *Walter Nernst*

A R Lang: photographs

Martin Levin and The Saturday Review: from *'Phoenix Nest', Saturday Review*

L Mackinnon: *Thermoelectric effect*

Macmillan Journals Ltd: from *Nature*

Mactier Publishing Corporation: from *EEE: The magazine of circuit design*

Mathematical Association of America: from *American Mathematical Monthly*

James E Miller and the publishers: from *American Scientist*

Museum of the History of Science, Oxford: photographs of Persian Astrolabe (IC 5)

Robert A Myers and the publishers: from *Physics Today*

National Physical Laboratory and 'Simplicius': from *NPL News*

Negretti and Zambra Ltd: from their *Encyclopaedic Catalogue*

New Statesman: poem

The New Yorker Magazine Inc and the author: *Perils of modern living* by H P Firth (copyright 1956)

The New Zealand Mathematics Magazine and the author: *The uses of fallacy* by Paul V Dunmore

The Niels Bohr Institute, L Rosenfeld and O R Frisch: from *Journal of Jocular Physics*

The Nonesuch Press Ltd: from *The Complete Works of Lewis Carroll*

North-Holland Publishing Co and the author: *The analysis of contemporary music using harmonious oscillator wave functions* by H J Lipkin

Jay M Pasachoff: poem

The Plessey Company Ltd: photograph of lead tin telluride crystals, taken on the Cambridge Stereoscan at their Allen Clark Research Centre

Prentice-Hall, Inc: from *A Stress Analysis of a Strapless Evening Gown*

Radio Corporation of America: from *Electronic Age*

Arthur Roberts: poem

Royal Greenwich Observatory: from *The Observatory*

The Royal Institution: Gillray's print '*Scientific Researches*'

The Royal Society and N Kurti: from *Biographical Memoirs of Fellows of the Royal Society*

George H Scherr: from *The Journal of Irreproducible Results*

Science Service Inc, Washington DC: from *Science News* (copyright 1968)

Scientific American Inc: drawing of Antikythera machine, from 'An Ancient Greek Computer' by Derek J de Solla Price. Copyright © 1959 by Scientific American Inc. All rights reserved.

Simon and Schuster Inc: from *The Space Child's Mother Goose* by Frederick Winsor and Marian Parry (copyright 1956, 1957, 1958)

Philip A Simpson: *Standards for inconsequential trivia*

Arthur H Snell: poem

Derek J de Solla Price: photograph of Antikythera machine

Springer-Verlag: from *Die Naturwissenschaften*

Norman Stone: *Face to face with metrication*

Studio International Publications Ltd and Jasia Reichardt: from *Cybernetic Serendipity exhibition catalogue*

J Sykes and S Chandrasekhar: *On the imperturbability of elevator operators*

Thames and Hudson Ltd: from *Science in the 19th Century*, editor René Taton

University of Toronto Press, North-Holland Publishing Co and the author: from *Elements of Conferencemanship* by D H Wilkinson

UNESCO: from *Impact of Science on Society*

Robert Weinstock: *Two classroom stories*

Dr P A Weiss and The Rockefeller University Press: *Life on Earth (by a Martian)*

Eugene Wigner: *Wolfgang Pauli*

WOR, New York: from *The Answer Man*

D A Wright and *The Worm Runner's Digest*: *A Theory of Ghosts*

206

Printed in the United States
by Baker & Taylor Publisher Services